Preface

What Descartes did was a good step. ...If I have seen further it is by standing on the shoulders of Giants.

— Isaac Newton [1676]

This book is designed to show how to use the software package Derive® [1] to help solve problems from calculus and related subjects. It is a companion to any of the textbooks used in calculus courses.

Derive is a computer algebra system (CAS) that provides capabilities to perform symbolic, graphic, and numeric manipulations. It is mostly menu driven, making it user-friendly and easy to use, which are necessary for an educational problem-solving tool. Its screen displays are nicely formatted and easy to read. In capable hands Derive is a powerful tool that can help solve problems encountered in calculus courses and in other subjects as well.

This manual is organized into four chapters. The first contains a general explanation of the basic capabilities and limitations of Derive. This section generally covers the symbolic, graphic, and numeric capabilities of the software in the areas of algebra, calculus, matrix algebra, and special functions. The basic keystrokes, menu commands, mode selection, plotting parameters, and screen displays are presented along with examples on

---

[1] Derive® is a registered trademark of Soft Warehouse, Inc., Honolulu, HI.

the use of Derive commands to perform simple calculus operations. The second chapter contains examples of solutions of problems similar to those in calculus textbooks using Derive to perform some of the manipulation, plotting, and analysis. Some of the problems are models from applications. Other problems deal directly with the mathematical concepts of topics in the course. The third chapter includes laboratory exercises which require the use of Derive to solve, explore, and analyze problems. Some of these problems are similar to those in the second chapter, while others introduce new concepts in both the mathematics and the software. These laboratory exercises lead the reader through the steps of problem solving. Sometimes these exercises involve conducting a computational experiment to determine an answer. Students should perform these laboratory exercises and experiments to develop their problem solving skills. The fourth chapter contains the answers to the exercises in the third chapter along with some suggestions and helpful hints on how to solve these problems.

In writing this manual, I have assumed that the reader either is already familiar with calculus or is concurrently studying the subject while using this manual. The object of this book is not to teach calculus, but to help in that endeavor.

I refer students who are studying or will be studying differential equations and instructors of courses on that subject to a similar manual about differential equations that I wrote. The book is entitled *Derive Manual for Differential Equations* and was published by Addison-Wesley in 1991.

It is hoped that this guide can contribute to better student understanding and performance in problem solving. It illustrates the power of Derive as a computational tool that enables students to become better problem-solvers.

Many people have helped make this book possible. I give special thanks to Frank Giordano for his encouragement and to Kathleen Snook and Fred Rickey for their thoughtful suggestions. I thank the creators of Derive, David Stoutemyer and Albert Rich, for their excellent software and suggestions. Finally, I give special thanks to Katie, Lisa, Dan, Kristin, and Sue for their support and understanding.

Chris Arney
West Point, NY
27 February 1991

Exploring CALCULUS
WITH DERIVE®

ISBN 0-201-52839-8

Copyright © 1992 by Addison-Wesley Publishing Company, Inc.

All rights reserved. No part of this publication may be reproduced, stored in a retrieval system, or transmitted, in any form or by any means, electronic, mechanical, photocopying, recording, or otherwise, without the prior written permission of the publisher. Printed in the United States of America.

3 4 5 6 7 8 9 10 AL 95949392

Notation and Technical Information

"What is the use of a book," thought Alice, "without pictures or conversations?"

— Lewis Carroll, *Alice in Wonderland* [1865]

This manual was typeset using LaTeX. This text processing system allows for special notation and fonts to be used to help identify the computer keystrokes and input commands. Boldface type is used to indicate the precise key on the keyboard to strike. This is especially helpful for identifying the special keys like **Enter**, **Ctrl**, **Alt**, and the function keys (**F1, F2, ..., F10**). The **Ctrl**, **Alt**, and **Shift** keys are often used in combination with another key and this is denoted by placing a hyphen between them and showing both keys in boldface (i.e. **Alt-e** or **Ctrl-Enter**).

The teletype font is used for the commands and menu selections for input to or output from the software (for example Author, Simplify, Help, Calculus, $\boxed{\texttt{LN(2z/3)}}$, $\boxed{\texttt{cos(xy-x)}}$, $\boxed{\texttt{DIF(e\^{}x+sin(x),x,2)}}$). When two or more commands are given in sequence they are separated by a space (i.e., `Declare Variable`). The in-line commands are usually in upper case to distinguish them from the menu selections and are displayed in a box for further emphasis when they are in the exact form to be entered into the computer. Derive is not sensitive to case unless requested to be, so actual input ordinarily may be made with upper-case or lower-case characters.

Sometimes italics are used to indicate a pseudo-command. For example, the selection of the Author menu command and the input of an expression can be denoted by *Authoring* the expression. This notation is not used often and only after the command is very familiar to the reader.

Words in uppercase letters and in regular typeface are used to indicate DOS file names. Derive uses utility files to store extra commands and functions, system states, and output expressions. These files are loaded whenever their commands are needed.

Version 2 of Derive is used in all the examples in this book. Earlier versions of the software lacked many of the utility files which play a significant role in solving many problems.

There are numerous figures throughout the manual showing actual screen images from Derive. Usually, the highlight is moved off the visual region of the screen so it will not interfere with the display. The screen images were produced using the CAPTURE program of Microsoft Word and plotted in Postscript format on a NEC Laser Printer. The computer that was used to run the software was a Zenith 248Z, which is compatible with an IBM-AT.

Contents

PREFACE		iii
NOTATION AND TECHNICAL INFORMATION		v
CHAPTER 1	**Getting Started**	1
	1.1 An Overview of Derive	2
	1.2 Menus and Commands	5
	1.3 Keystroking and Entering Expressions	11
	1.4 Screen Displays and Windows	13
	1.5 On-Line Help and System State	16
	1.6 Symbolic Algebra	18
	1.7 Calculus	20
	1.8 Matrix Algebra	23
	1.9 Plotting	27
	1.10 Numerical and Functional Approximations	34
	1.11 Programming	36
	1.12 Demonstration of Commands and Capabilities	40
	1.13 Utility Files for Calculus	45
	1.14 Differential Equations	51
	1.15 Limitations	54
	1.16 Recommendations	55
CHAPTER 2	**Examples**	57
	2.1 Limits and Continuity	57

2.2	Curve Sketching	63
2.3	Derivatives	69
2.4	Curvature	76
2.5	Root Finding	81
2.6	Integration	85
2.7	Arc Length	90
2.8	Projectile Motion	92
2.9	Radioactive Decay	96
2.10	Area and Volume	99
2.11	Polar Equations	106
2.12	Building a Pool	110
2.13	Taylor Polynomial	114
2.14	Separable Differential Equation	118
2.15	Exact Differential Equation	123
2.16	Drug Doses	126
2.17	Electrical Circuit	128
2.18	Numerical Solution of Differential Equations	134
2.19	Linear Algebra	137
2.20	Series	142

CHAPTER 3   Exercises   145

3.1	Limits and continuity	146
3.2	Function Race	146
3.3	Rectilinear Motion of a Robot	147
3.4	Area Between Functions	148
3.5	Mass, Moments, and Centroids	148
3.6	Price of Electricity	149
3.7	Building a Dam	150
3.8	Deflection of a Beam	151
3.9	Taylor Series	152
3.10	Bouncing Ball	152
3.11	Measuring Earthquakes	153
3.12	Population Growth	154
3.13	Newton's Law of Cooling	154
3.14	Solving Systems of Linear Equations	155
3.15	Solving Systems of Nonlinear Equations	156
3.16	Minimizing Heat Loss	156
3.17	Volume	157
3.18	Oil Spill	157
3.19	Inventory and Pricing	158
3.20	Automobile Suspension System	159

OTHER READING   161

INDEX   163

1
Getting Started

Today we render unto the computer what is the computer's, and unto analysis what is analysis', we can think in terms of general principle, and appraise methods in terms of how they work ....

— Peter Lax [1989]

This chapter describes the fundamentals of using Derive to solve mathematics problems and is a good preparation before reading the example problems solved in Chapter 2. Since calculus involves many areas of mathematics, sections are included on the use of Derive in the subject areas of matrix algebra, complex variables, and differential equations. This chapter also includes the fundamental commands of the software in symbolic algebra, plotting, and numerical approximations. One section demonstrates the use of Derive commands to solve simple drill problems in calculus. However, neither this manual nor this first chapter are complete reference manuals for Derive. This book is not intended to replace the *Derive User Manual* that comes with the software. In fact, at times this manual will refer you to the *Derive User Manual* for more information.

Beginning users of Derive should definitely spend some time in this chapter before tackling the examples in Chapter 2. The material in this chapter is designed to be a helpful reference for reading Chapter 2, solving problems from Chapter 3, and using Derive to solve problems in any area of mathematics.

1.1 An Overview of Derive

Making mathematics more exciting and enjoyable is the driving force behind the development of Derive. The system is destined to eliminate the drudgery of performing long tedious mathematical calculations.

—Derive User Manual [1990]

It is important to realize that the Derive software was not specially designed to solve problems in just one field, such as calculus. Derive is a computer algebra system (CAS) whose function is more general than any one subject or topic. It has capabilities to perform symbolic, numeric, and graphic operations. There are other packages such as Calculus Toolkit,[1] CALCULUS PAD,[2] Calculus,[3] Mathematics Plotting Package,[4] Exploring Calculus,[5] Eureka: The Solver,[6] and MathCAD[7] that help solve some special kinds of calculus problems and demonstrate the concepts of calculus. Other packages like Phaser,[8] MDEP,[9] MacMath,[10] and Differential Equations Graphics Package[11] are designed for the expressed purpose of analyzing and numerically solving differential equations. Software programs like LinTek,[12] MAX,[13] MATLAB,[14] LINDO,[15] and the Linear Algebra Toolkit[16] help solve matrix and linear algebra problems. All these packages possess some of the capabilities of a CAS, but usually lack the versatility of CAS software.

---

[1] Ross Finney, et al., Addison-Wesley.
[2] I. Bell, et al., Brooks-Cole.
[3] John Kemeny, TrueBASIC.
[4] H. Penn, United States Naval Academy.
[5] John Fraleigh and Lewis Pakula, Addison-Wesley.
[6] Borland.
[7] Mathsoft; student version available from Addison-Wesley.
[8] H. Kocak, Springer-Verlag.
[9] J.L. Buchanan, United States Naval Academy.
[10] J.H. Hubbard and B.H. West, MacMath.
[11] Sheldon Gordon, MatheGraphics.
[12] John Fraleigh, Addison-Wesley.
[13] E.A. Herman and C.H. Jepsen, Brooks-Cole.
[14] Little and Moler, The Math Works.
[15] Scientific Press.
[16] C. Wilde, Addison-Wesley.

Derive performs many of the mathematical steps and approximations involving algebra, calculus, trigonometry, number theory, and plotting necessary to solve, analyze, and study many types of problems. Other CAS software packages similar to Derive for personal computers are Maple,[17] Mathematica,[18] muMath,[19] Macsyma,[20] Reduce,[21] Mathematics Exploration Toolkit,[22], and Theorist.[23]

The following figure shows part of a sample Derive screen display from a problem-solving session showing the general layout of Derive's user interface.

```
1:   SIN (2 x + π)
           b
2:   ∫    (e^x - x^2 + 4 x) dx
           a

3:   LN (x + y)

COMMAND: aUthor Build Calculus Declare Expand Factor Help Jump soLve Manage
              Options Plot Quit Remove Simplify Transfer moVe Window approX
Enter option
User                    D:1.1              Free:100%              Derive Algebra
```

Derive screen showing three working expressions in the work area (expression # 3 is highlighted) and the main menu.

---

[17] Waterloo Maple Software.
[18] Wolfram Research.
[19] The predecessor of Derive.
[20] Symbolics.
[21] Northwest Computer Algorithms.
[22] WICAT Systems.
[23] Prescience.

The screen is organized into several sections each with an important function. The top part contains the working expressions. This sample of this section of the Derive screen shows expressions 1-3.

$$1: \quad \text{SIN}(2x + \pi)$$

$$2: \quad \int_a^b (\hat{e}^x - x^2 + 4x)\, dx$$

$$3: \quad \text{LN}(x + y)$$

Three algebraic working expressions (numbered 1-3) in the top part of a Derive screen.

This section can also include plots and can be partitioned into subsections by the user. The screen below shows three windows, one containing an algebraic expression, another window containing a 2-dimensional plot, and the third has a 3-dimensional plot. Windowing and plotting are discussed in Sections 1.4 and 1.10, respectively.

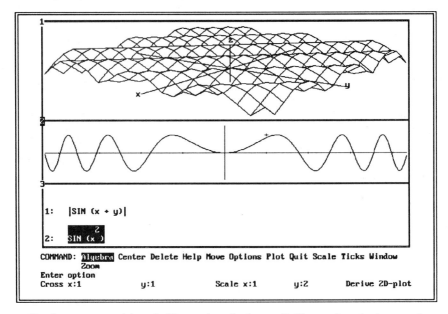

Derive screen with a 3-dimensional plot, a 2-dimensional plot, and algebraic expressions.

1.2 Menus and Commands

Computers change not only how mathematics is practiced, but also how mathematicians think.

— Lynn Arthur Steen [1990]

The section below the expressions on the Derive screen contains the menu. Commands are either typed directly into the work area using the Author menu command or are selected from the menu in one of two ways. The first way is merely to type the letter associated with the menu command (the upper case letter in the display of the command, which is not always the first letter of the word). The second way is to move the highlight in the command line using **Space, Tab, Back Space,** or **Shift-Tab** keys and pressing the **Enter** key when the desired menu command is highlighted. The **Esc** key is used to abort a command or menu without making any changes. The main menu and the Author command are the users' usual interface with Derive.

```
COMMAND: Author Build Calculus Declare Expand Factor Help Jump soLve Manage
         Options Plot Quit Remove Simplify Transfer moVe Window approX
Enter option
                                    Free:100%                Derive Algebra
```

Derive's familiar main menu and its command selections.

Many of the commands are executed directly from the main menu. These executable commands include Author, Build, Expand, Factor, Help, Jump, soLve, Quit, Remove, Simplify, approX, and moVe. The following table gives a short description of each of these executable main menu commands.

Command	Function
Author	Allows for the input of expressions into the algebra window or work area.
Build	Provides for the building of an expression from previous expressions.
Expand	Performs the algebraic expansion of an expression.
Factor	Factors an expression. If the expression is a polynomial, factors it into the lowest degree terms meeting the criterion established with the mode portion of the command. If the expression is a number, expresses the number in terms of its prime factors.
Help	Provides help on the syntax and function of commands (see Section 1.5).
Jump	Moves the highlight to a given expression number. This is a good way to move to an earlier expression in the work area.
soLve	Solves an equation for the desired variable.
Quit	Stops the Derive program. Control returns to DOS.
Remove	Removes expressions from the work area.
Simplify	Simplifies an expression, equation, or a Derive command in the work area. This command is used frequently.
moVe	Rearranges expressions in the work area.
approX	Approximates the value of an expression. Precision is set with the `Options Precision` command.

Main menu commands and their functions.

Some other commands merely open submenus which contain additional commands. The submenu commands also can be one of these two kinds. Therefore, the menu structure is much like a tree with several branches. The user must learn what branch to follow to obtain the needed executable command or learn the proper command name so it can be typed directly into the work area. Some commands can be entered both ways, while others are available only through the menu or only through a typed command.

The following figure shows a schematic of the tree-like menu structure. Only the most common commands are included in the figure.

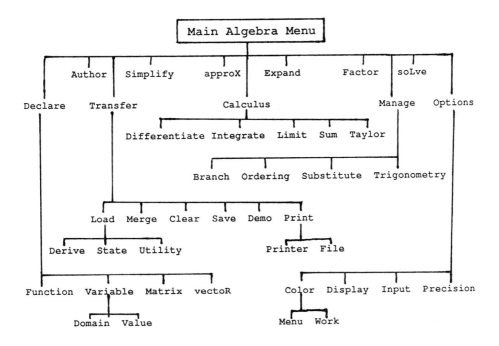

Most of the important submenus and their associated commands are explained in the following paragraphs and tables. Other submenus are discussed in later sections of the manual.

The set of commands in the submenu Declare are Function, Variable, Matrix, and vectoR. These are very useful and powerful command and careful thought in their use is often needed to solve problems correctly. These commands will be used in some of the examples and may be necessary in some of the laboratory exercises. Not only can you declare these mathematical structures (matrix, vector, etc.), but you can also establish important domain information on variables using the Declare Variable Domain command and the components in the structures can be directly entered. The following table describes the function of these commands.

Command	Function
Function	Declares a function and optionally defines it.
Variable	Assigns the domain of a variable (options include Positive, Nonnegative, Real, Complex, and Interval) or assigns the variable a value.
Matrix	Establishes the dimensions of a matrix and provides for the entry of elements in the matrix.
vectoR	Establishes the dimension of a vector and provides for the entry of the elements in the vector.

Declare menu commands and their functions.

The Manage submenu contains the commands Branch, Exponential, Logarithm, Ordering, Substitute, and Trigonometry. Probably, the most important and frequently used command in this list is the Manage Substitute command. This command substitutes values for variables or subexpressions of an expression. The location of the Substitute command in the Manage submenu is worth remembering. More information on these commands is found in the *Derive User Manual*.

The Options submenu sets up different modes of operation for Derive. Some of the Options commands are usually used before any evaluations are done. Other commands may be performed during the course of problem solving. The commands in this submenu are Color, Display, Execute, Input, Mute, Notation, Precision, and Radix. The following table provides a short description of these commands. It is important to set the proper parameters for your computer using the Options Display command. These settings can be saved via the Transfer Save State command.

Command	Function
Color	Changes the color of the work area, menu, or plotting curves.
Display	Sets Mode to Text or Graphics and establishes the Resolution and Adapter.
Execute	Allows for the execution of DOS commands.
Input	Establishes the input mode for variables as single letters or words.
Mute	Sets the error beep on or off.
Notation	Sets the style of notation for numerical output and the number of digits displayed.
Precision	Sets the computational precision mode and the digits of accuracy for the Approximate mode.
Radix	Sets the radix base for number input and output.

Options menu commands and their functions.

The Transfer submenu provides for input and output in Derive. The subcommands and their functions are given in the following table. This menu is used to load utility files, to save files, to run demonstration files, to save and to load system state files, and to obtain hard-copy print outs of the expressions in the work area.

Command	Function
Merge	Adds the expressions stored in a specified .MTH file to those in the work area.
Clear	Deletes all the expressions from the work area.
Demo	Loads and simplifies expressions from a .MTH file one expression at a time.
Load	Loads all the expressions in a utility file, user defined Derive (.MTH) file, or State (.INI) file into the working area or to set up Derive parameters.
Save	Saves the expressions in the working area to a file. Format options include Derive (.MTH), FORTRAN, Pascal, or Basic or to save the current State. The State option saves the current state of Derive parameters to a .INI file for reloading.
Print	Prints the expressions in the work area on a printer or to a file. Allows for the setting of several printing options.

Transfer menu commands and their functions.

There are some commands (usually in the form of operators or functions) in Derive that are not reached through any menu. These commands and functions must be typed into the workspace using the Author command. Some of the more useful and common commands not in a menu are CURL, FIT, GRAD, LAPLACIAN, VECTOR_POTENTIAL, STDEV, RMS, ROW_REDUCE, EIGENVALUES, DET, DIMENSION, PRODUCT, POTENTIAL, PHASE, IM, and REAL. Explanations of these operations are found in the *Derive User Manual* and are available on-line through the Help menu. Some of these operations will be used in the Examples in Section 2.

1.3 Keystroking and Entering Expressions

Derive not only has to be a tireless, powerful, and knowledgeable mathematical assistant, it must be an easy, natural, and convenient tool.

— *Derive User Manual* [1990]

As discussed in Section 1.2, the keystrokes to select menu commands in Derive are straight forward. Just move the highlight by hitting the space bar and press **Enter** when the desired command is highlighted or merely type the letter that is capitalized in the desired command. However, there are many other keystrokes to learn in order to enter and edit expressions when using the Author and other similar commands. Keyboard skill is essential and a prerequisite for efficient use of the software. Learning to use the editing features can reduce considerably the amount of retyping of expressions.

One feature is the ability to input comments into the work area between expressions. Comments are entered by putting quotes around the line of input when entering an expression.

When entering and editing an expression using the Author command, the **BackSpace** deletes the previous character (left of the cursor). **Ctrl-S** moves the cursor one character to the left without erasing the character, and **Ctrl-D** moves the cursor to the right. **Ctrl-A** moves the cursor a group of characters (or token) to the left, and **Ctrl-F** moves the cursor a token to the right. **Ctrl-Q-S** moves the cursor to the left end of the line, and **Ctrl-Q-D** moves the cursor to the far right end. **Del** deletes the character at the cursor. **Ctrl-T** deletes a token at a time, and **Ctrl-Y** deletes the entire line.

Ins toggles the typing modes between inserting between existing characters and overwriting any existing characters. When the insert typing mode is in effect, the word Insert appears on the status line below the menu.

The direction keys are used to highlight and obtain subexpressions (or entire expressions) from previous expressions in the working area. These subexpressions can be expanded or assembled into new expressions. The ↑ and ↓ move the highlight one expression at a time. **PgUp** and **PgDn** move the highlight several expressions at once. **Home** moves to the first expression, and **End** highlights the last expression in the working area. The other direction keys, ← and →, move the highlight through the individual tokens (or subexpressions) of an expression. Even finer highlighting is possible (See the Users Manual). The **F3** function key inserts whatever is highlighted into the author line. The **F4** key does the same and also automatically places parentheses around the expression. Similar things can be done using the Build command (See Section 1.2). The easiest way to refer to previous expressions is by their number in the work area. Just use the # symbol preceding the number of the expression to obtain that expression. For example, authoring #4/#6, inserts expression number 4 divided by expression number 6 into the work area.

The **Enter** key completes the expression and adds it to the work area. **Ctrl-Enter** also does this and automatically *Simplifies* the expression.

There are simple keystrokes to enter the common special characters. The symbolic entry for the irrational numbers π and e are **Alt-p** and **Alt-e**, respectively. π can also be entered by typing pi. The imaginary unit i is entered by **Alt-i**. The square root symbol $\sqrt{}$ is entered by **Alt-q** or by using the function SQRT—for example, SQRT(x).

There are two input modes for variable names—character or word. The mode is set with the Options Input command. In the Character mode, each letter represents a variable. In the Word mode, entire words are used as variables. In Word mode, the words representing variables must be separated by a space or operator or appropriate punctuation, such as a parenthesis. The Options Input command also provides the mechanism to put Derive into case sensitive (upper case letters different from lower case) or case insensitive modes.

Some Greek letters can be used as variables and entered using the **Alt** key and a letter key or by typing their Latin names. The following chart provides the keys to use to obtain Greek letters:

Greek Letter	Key
α	Alt-a
β	Alt-b
γ	Alt-g
δ	Alt-d
ϵ	Alt-n
θ	Alt-h
μ	Alt-m
π	Alt-p
σ	Alt-s
τ	Alt-t
ϕ	Alt-f
ω	Alt-o

Keystrokes to enter Greek letters.

1.4 Screen Displays and Windows

My greatest hope is that it pleases those who have at heart the development of science and that it proposes solutions that they have been looking for or at least opens the way for new investigations.

— Carl Friedrich Gauss, *Disquisitiones Arithmeticae* [1801]

Derive allows the screen to contain more than one window. These windows can split the screen in a variety of ways or overlay one another. Windowing can be used for several purposes. Separate windows can contain different problem solutions, different developments of the same problem, different perspectives or scales of the same plot, or maintain important results in one window while working in another. Windows can be especially helpful in showing symbolic computations, numeric computations, and graphics all on the same screen in three or more different windows.

The Window command is called from the main menu. The Window menu is shown here:

```
WINDOW: Close Designate Flip Goto Next Open Previous Split
Enter option
                                    Free:100%        Derive Algebra
```

Windowing is a very important and powerful feature of Derive. Be sure to use windowing when it is helpful. The following table describes the use of the windowing commands.

Command	Function
Close	Closes the active window.
Designate	Designates the active window 1 of 3 types (Algebra, 2D plot, 3D plot).
Flip	Flips or rotates between overlaid windows (the windows are not numbered separately).
Goto	Activates and moves control to the designated window.
Next	Activates and moves control to the next window in sequence.
Open	Opens a new window and designates it 1 of 3 types (Algebra, 2D plot, 3D plot).
Previous	Activates and moves control to the previous window in sequence.
Split	Splits the current active window to form a new window. The split can be made vertically or horizontally and can be established in different sizes.

Window menu commands and their functions.

The following example shows three horizontal windows. The top window contains the symbolic form of the function, the middle window shows the global behavior of this function, and the bottom window gives a smaller scale perspective or local behavior near the origin. The number on the active window is highlighted (window #3 in the following figure).

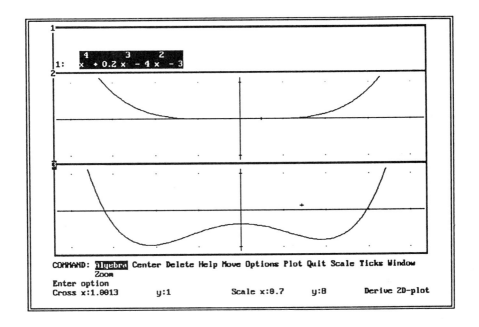

There is a convenient keystroke that saves time when moving between windows. The **F1** key is equivalent to the Window Next command and provides easy and rapid movement through the existing windows.

1.5 On-Line Help and System State

It is better to know some of the questions than all of the answers.

— James Thurber [1945]

The Help command displays the following menu:

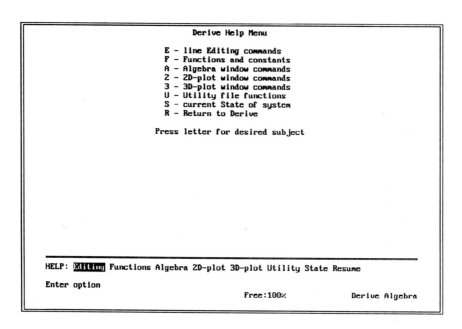

To obtain on-screen help on any of these subjects, press the appropriate letter in this menu. Helpful information is provided along with the section number in the *Derive User Manual* which contains more information. There is plenty of information available, so don't be hesitant to use this feature.

The Help, S option displays the current system state. There are several screens of information available through this command. The first screen provides the basic mode settings and the second screen contains mostly information on the plotting parameters and output formats. Information is provided about the status of the display modes set with the Options Display menu commands (Text or Graphics, resolution, type of graphics adapter—CGA, EGA, VGA, etc.) by use of a numerical code. All of these modes can be saved and established in the .INI file. The system status for the set-up used to solve the examples and exercises in this manual are provided in the following output.

```
                    System Control Settings

Algebra window numerical settings
   Precision mode:  Exact
   Precision digits:  6
   Notation style:  Rational
   Notation digits:  6
   Input radix base:  10
   Output radix base:  10
   Input mode:  Character
   Case mode:  Insensitive

Algebra window simplification settings
   Branch selection:  Principal
   Logarithms:  Auto
   Exponentials:  Collect
   Trig functions:  Auto
   Trig powers toward:  Auto
```

```
                    System Control Settings

Plot window plotting settings
   Rows per tick mark:  4
   Columns per tick mark:  9
   Plotting accuracy:  7
   Coordinate type:  Rectangular
   Auto change color:  No
   Axes color:  15
   Cross color:  15
   3D plots top color:  6
   3D plots bottom color:  5

Printer settings
   Page length:  66
   Page width:  80
   Top margin:  0
   Bottom margin:  0
   Left margin:  8
   Right margin:  3
```

```
                    System Control Settings

Window color settings
   Frame color:  15
   Option color:  15
   Prompt color:  15
   Status color:  15
   Menu background color:  0
   Border color:  0
   Work color:  15
   Work background color:  0

Display mode settings
   Video mode:  18
   Character set:  Extended
```

These state parameters can be saved into and reloaded from an .INI file using the `Transfer Save State` and `Transfer Load State` commands, respectively. Derive starts in the state set in the file DERIVE.INI. If a different initial state is desired, save that state in the file DERIVE.INI with the `Transfer Save State` command and that will become the initial state of the software. Different states can be saved in other .INI files with the `Transfer Load State` command and loaded whenever needed. See Section 1.2 or the *Derive User Manual* for more information about the system state and refer to Section 1.16 of this book for recommendations on establishing proper mode settings for your own DERIVE.INI file.

1.6 Symbolic Algebra

The performance of the computer is to be judged by the contribution which it will make in solving problems of new types and developing new methods.

— John von Neumann [1946]

Derive has the capability to do many algebraic manipulations using the `Simplify`, `soLve`, `Manage`, `Expand`, and `Factor` commands. See Section 1.2 or the *User Manual* for explanation of these commands. Derive can work with equations, expressions, and inequalities. It also can do operations with ∞. Enter ∞ by typing `inf`.

Derive assumes the default declaration of `Real` for all variables, unless over-ridden by the `Declare Variable Domain` command. In a similar manner the `Declare Variable Value` command creates user-defined variables and optionally gives them values. The constant number e is entered by **Alt-e** and displayed as ê, and the constant π is entered as `pi` or **Alt-p**. Three files are available to the Derive user with many useful variables already declared. These files are ENGLISH.MTH, METRIC.MTH, and PHYSICAL.MTH. These files can be loaded when needed using the `Transfer Load Derive` or `Transfer Load Utility` commands.

`Declare Function` is a very powerful feature in Derive. It creates user-defined functions which also can be used as functional operators.

Some of the common algebraic and trigonometric functions in Derive are provided in the following table.

Command	Function
EXP(z)	Exponential of z, displayed as $\hat{e}^z$.
SQRT(z)	$z^{1/2}$. Takes into account the domain of the variable and the branch setting.
LN(z), LOG(z,w)	Principal natural logarithm and log to the base w.
PI, DEG	Constants π and $\pi/180$.
SIN(z), COS(z), TAN(z) COT(z), SEC(z), CSC(z)	Trigonometric functions.
ASIN(z), ACOS(z), ATAN(z) ACOT(z), ASEC(z), ACSC(z)	Inverse trigonometric functions.
SINH(z), COSH(z), TANH(z) COTH(z), SECH(z), CSCH(z)	Hyperbolic trigonometric functions. Simplify to the exponential equivalents.
ASINH(z), ACOSH(z), ATANH(z) ACOTH(z), ASECH(z), ACSCH(z)	Inverse hyperbolic trig functions. Simplifies to the logarithmic equivalents.
ABS(x), SIGN(x)	$\lvert x \rvert$ and sign of x, respectively.
MAX($x_1, ..., x_n$), MIN($x_1, ..., x_n$)	Max and min of all the arguments.
STEP(x)	1 if $x > 0$, 0 if $x < 0$. Provides for the definition of piecewise functions.

Common algebra and trigonometry functions.

The trigonometric manipulation in Derive is controlled with the command **Manage Trigonometry**. There are several possible combinations of ways to manage and simplify the trigonometric functions. Derive uses radians for angular measure in all its trigonometric functions.

Derive can perform complex arithmetic and can handle complex variables and complex-valued functions. Variables are declared complex using the `Declare Variable Domain` command. The imaginary unit i is entered using **#i** or **Alt-i** and displayed as $\hat{i}$. Some of the common functions involving complex numbers are given in the following table.

Command	Function
`ABS(z)`	$\|x + \hat{i}y\|$
`SIGN(z)`	Finds the point of unit magnitude with the same phase angle as z.
`RE(z)`	Computes the real part of z, (x).
`IM(z)`	Computes the imaginary part of z, (y).
`CONJ(z)`	Finds the complex conjugate of z.
`PHASE(z)`	Computes the principal phase angle of z.

Common functions involving Complex Variables.

There are many other functions in Derive to perform operations in probability, statistics, and finance. See the *Derive User Manual* for descriptions of these functions.

1.7 Calculus

Calculus is first of all a language, not only for scientists, but also for economists and social scientists.

—Lynn Arthur Steen [1990]

Derive performs the Calculus operations of limits, differentiation, antidifferentiation, integration, Taylor polynomial approximations, series summation, and products. These operations can be done through the `Calculus` submenu or with in-line commands. Several items about these operations are worth mentioning: i) antidifferentiation does not automatically produce an arbitrary constant, ii) definite integration in the `Approximate` mode is performed with an adaptive variant of Simpson's rule, and iii) definite integrals in exact mode are evaluated using one-sided limits of an antiderivative, so it is your responsibility to account for any internal singularities.

When the Calculus submenu is selected, a new menu appears at the bottom of the screen. The executable commands in this submenu are Differentiate, Integrate, Limit, Product, Sum, and Taylor. These commands are used in many of the examples in Chapter 2. The following table provides descriptions of these commands.

Command	Function
Differentiate	Finds the derivative (regular or partial) of the highlighted expression to any order.
Integrate	Finds either the antiderivative or the definite integral of a function, depending on whether limits of integration are entered.
Limit	Finds the limit of an expression as a variable approaches a value. The approach can be from the left, right, or both.
Product	Finds the definite product or antiquotient of an expression. The index variable is required.
Sum	Finds the definite sum or antidifference of an expression. The index variable is required.
Taylor	Finds the Taylor polynomial approximation to an expression. The expansion variable, expansion point, and order of the polynomial are required.

Calculus menu commands and their functions.

These calculus menu commands also have equivalent in-line commands. The following table on the next page explains the format and use of these in-line commands. The in-line commands must be entered along with their arguments using the Author menu command; the Simplify command then executes the operation.

Command	Function
DIF(u,x,n)	Finds the nth-order derivative of u with respect to x.
INT(u,x)	Finds the antiderivative of u with respect to x.
INT(u,x,a,b)	Finds the definite integral of u with respect to x from a to b.
LIM(u,x,a,0)	Finds the limit of u as x approaches a from both left and right sides.
LIM(u,x,a,1)	Finds the limit of u as x approaches a from the right (above).
LIM(u,x,a,-1)	Finds the limit of u as x approaches a from the left (below).
PRODUCT(u,n)	Finds the antiquotient of u with respect to n.
PRODUCT(u,n,k,m)	Finds the definite product of u as n goes from k to m.
SUM(u,n)	Finds the antidifference of u with respect to n.
SUM(u,n,k,m)	Finds the definite sum of u as n goes from k to m.
TAYLOR(u,x,a,n)	Finds the Taylor polynomial approximation of order n to u about the point $x = a$

Calculus in-line commands and their functions.

There are three utility files that contain additional Calculus-related functions: DIF_APPS.MTH, INT_APPS.MTH, and MISC.MTH. See Section 1.13 for more information about these utility files.

Derive also has several vector calculus functions. The default for the vector function is 3-dimensional rectangular coordinates x, y, and z. However, the utility file VECTOR.MTH can help perform coordinate transformation and produce operation in other coordinate systems. See the *Derive User Manual* for the description of this file and its use. The following table provides descriptions of some of the vector calculus commands.

Command	Function
GRAD(f(x,y,z))	Finds the gradient of an expression.
DIV([u(x,y,z),v,w])	Finds the divergence of a vector function.
LAPLACIAN(f(x,y,z))	Finds the divergence of the gradient of a function.
CURL([u,v,w])	Computes the curl of a vector.
POTENTIAL([u,v,w])	Calculates the scalar potential of a vector.
VECTOR_POTENTIAL ([u,v,w])	Calculates the vector potential of a vector.

Common vector calculus functions.

1.8 Matrix Algebra

It will be seen that matrices comport themselves as single quantities.

— Arthur Cayley [1858]

Solving systems of linear equations and performing matrix and vector operations are important and frequent tasks in solving problems involving systems of differential equations. Derive has capabilities to help the user perform these operations.

Vectors can be entered in several ways. Four common ways are: i) Author an expression of the form $[x_1, x_2, \ldots, x_n]$, ii) use the `Declare Vector` command which prompts the user for the size and elements of the vector, iii) use the `VECTOR(f(k), k, n)` command to obtain vector elements $f(k), k = 1, 2, \ldots, n$, or iv) use the `ITERATES` command to obtain vector elements obtained from a recursive sequence. For example, using the method in item (iii) above, Author `VECTOR(k^3,k,4)` and `Simplify` to obtain to [1,8,27,64]. Other stepping parameters for the indices can be used in the VECTOR command. See the *Derive User Manual* for details about the parameters of this command.

Matrices are entered in similar fashions with vectors acting as rows of the matrix. Author

`[[1,2,3],[a,b,c],[x,sinx,6!]]`,

which results in the matrix

$$1: \begin{bmatrix} 1 & 2 & 3 \\ a & b & c \\ x & \text{SIN}(x) & 6! \end{bmatrix}$$

Similarly, the `Declare Matrix` command prompts for the dimensions and elements of a matrix. Nested calls to the VECTOR command can produce a matrix with function evaluations for values of the elements. Additionally, the identity matrix of n dimensions is produced with the `IDENTITY_MATRIX(n)` command.

The . (period or dot) is the operator used to perform dot products and matrix multiplication between vectors and matrices. The dimensions of the vectors and matrices must be consistent for this operator to perform. The ` key is used to transpose a matrix or vector. This is not the apostrophe key. The ` key is usually found with the tilde rather than the quotation mark. The ^ key is used to produce powers of matrices. The ^(-1) symbol computes the matrix inverse. Descriptions of the most common matrix operations are provided in the following table.

Command	Function
ELEMENT(m,i,j)	Extracts element (i,j) from a vector or matrix m.
CROSS(v,w)	Calculates the cross product of 2 vectors $(v \times w)$.
DIMENSION(m)	Determines the number of rows in matrix m.
OUTER(v,w)	Computes the outer product of vectors v and w.
DET(m)	Calculates the value of the determinant of matrix m.
TRACE(m)	Sums the diagonal elements of matrix m.
ROW_REDUCE(m)	Computes the row echelon form of matrix m.
CHARPOLY(m)	Produces the characteristic polynomial of matrix m.
EIGENVALUES(m)	Determines the eigenvalues of matrix m by finding the roots of the characteristic polynomial.

Matrix manipulation commands and their functions.

To solve a system of linear equations, the equations are entered as elements of a vector and the soLve command is issued. If there are more variables than equations, the system prompts the user for the variables to solve for in the output. If the system is singular, the solution will either contain arbitrary values (@1, @2, etc.) or will display the message: No solutions found, depending on whether or not the system is consistent.

Examples of input expressions and results for three linear systems are provided. The input for the first example is to select the Author command and enter

[x+y-0.4z=10,y-15z=-0.51,-2x-y+z=1].

The solution is found by executing the menu command soLve. The resulting solution is shown in expression #2 of the following figure.

The second example (expressions #3 and #4) is similar to the first, except as shown in expression #4 of the display screen, the output contains @1 which signifies an arbitrary constant.

The third example (expressions #5 and #6) contains four variables in the three equations, so Derive queries the user for the variables to be solved for when the soLve command is issued. In this case a, b, and c are the solve variables given. The solution is given in expression #6 in terms of the variable d.

```
1:   [x + y - 0.4 z = 10, y - 15 z = -0.51, - 2 x - y + z = 1]

2:   [x = - 77147/7600, y = 157449/7600, z = 2151/1520]

3:   [6 x + 3 y - z, - 2 x - 11 y + z, 7 x - 14 y]

4:   [x = @1, y = @1/2, z = 15 @1/2]

5:   [a + b + c = 3 d, 16 b - 2 d = 15, a - b - 3 c = 1]

6:   [a = 7 (10 d - 1)/32, b = 2 d + 15/16, c = 22 d - 23/32]
```

The ROW_REDUCE command can also be used to solve systems of linear equations. This is a very powerful computational tool in matrix algebra. The operation is performed upon execution of the Simplify command. The ROW_REDUCE command and its computed output to solve the first example are shown:

```
                  ⎡  1    1  -0.4   10  ⎤
7:   ROW_REDUCE   ⎢  0    1   -15  -0.51⎥
                  ⎣ -2   -1    1     1  ⎦

         ⎡ 1  0  0   - 77147/7600 ⎤
8:       ⎢ 0  1  0     157449/7600⎥
         ⎣ 0  0  1     2151/1520  ⎦
```

The utility file VECTOR.MTH contains commands for matrix manipulation. Some of the most powerful commands are presented in the following table.

Command	Function
PIVOT(A,i,j)	Performs one pivoting step in row reducing matrix A.
COFACTOR(A,i,j)	Finds the cofactor of element (i,j) of the matrix A.
EXACT_EIGENVECTOR(A,w)	Finds the eigenvector of matrix A for the exact eigenvalue w. Best used for matrices up to 3x3 in size.
APPROX_EIGENVECTOR(A,w)	Finds the eigenvector of matrix A for the approximate eigenvalue w. Best used for 4x4 and larger matrices.
JACOBIAN(u,θ)	Finds the Jacobian matrix of the system of equations $x = u(\theta_1, \theta_2, ...)$.

Some of the matrix manipulation commands in VECTOR.MTH and their functions.

Other commands in this file compute outer products; append vectors; and delete, swap, scale, and subtract vector and matrix elements. See the *Derive User Manual* for descriptions of these commands.

1.9 Plotting

All the pictures that science now draws of nature ... are mathematical pictures.

— Sir James Hopwood Jeans [1930]

Derive has the capability to do both 2- and 3-dimensional plotting. Both of these capabilities are available through the Plot command.

Before the plotting commands are discussed, another important command must be reviewed. Derive has several display modes, so it is important to set the correct mode for your hardware and computational task. The command used to set modes is `Options Display`. If your computer and screen have graphics modes, set the Mode to Graphics when plotting. The **F5** key switches to the previous display mode so it is handy to use to switch back and forth between text and graphics modes.

The 2-dimensional `Plot` screen and menu are as follows:

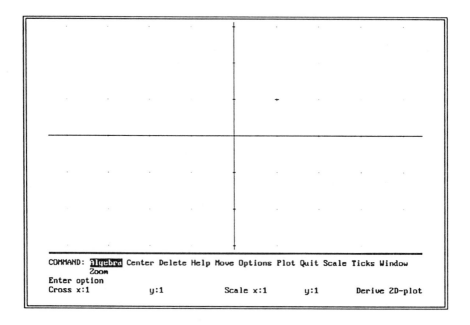

To produce a 2-dimensional plot in rectangular coordinates, the following procedure is used: i) highlight the expression to plot (it must be in the form $y = f(x)$ or just $f(x)$, but the variable names don't matter), ii) call the `Plot` menu, and iii) execute the `Plot` command.

If three or more functions need to be plotted on the same axes, they can be assembled as components of a vector function. Then all the functions in the vector will be plotted on the same axis at one time.

This produces a plot using the current plotting parameters. The plotting parameters are controlled with additional commands in the `Plot` menu. The following table gives a short explanation of these `Plot` commands.

Command	Function
`Algebra`	Changes control back to the algebra screen.
`Center`	Positions the center of the plot at the location of the movable cross.
`Delete`	Deletes functions from the plot list.
`Move`	Moves the cross to the specified coordinates.
`Options`	Sets parameters for `Accuracy`, `Color`, `Display`, and `Type`.
`Plot`	Plots the functions in the plot list and the highlighted function.
`Scale`	Sets plotting axes scales to exact values.
`Ticks`	Establishes the distances between tick marks on the axes. This controls the aspect ratio of the plot.
`Window`	Opens a new window. This is the same as the `Window` command in the Algebra window.
`Zoom`	Automatically changes the plot scale by a fixed amount. Zoom is made `In` or `Out` and in the x-, y-, or both directions.
`Quit`	Stops execution of the Derive program.

2-Dimensional `Plot` menu commands and their functions.

The following example shows a plot of the function $y = f(x)$ defined by
$$y = x + 2 \sin x$$
This function is entered with the `Author` command and by typing `x+2sinx`.

The `Plot` menu is selected to open the 2-D plotting window. Then the default plot parameter `scale` is changed to 4 in both the x and y directions. Finally, the `Plot` command is issued to produce the following plot:

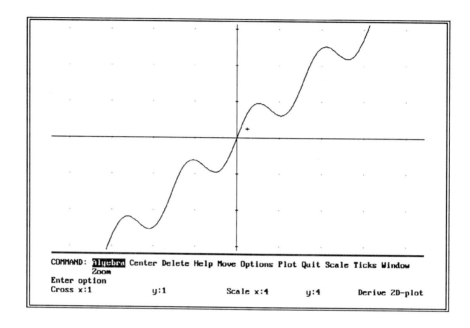

There are several helpful keystrokes and tricks to know while plotting. The direction arrow keys move the small cross around the plot screen. The coordinates of the cross are displayed in the status line at the bottom of the screen. This cross can be handy in finding approximations to many values of the plotted function. The **F9** and **F10** keys are equivalent to the Zoom In and Zoom Out commands, respectively. Derive will plot three or more functions at the same time when they are entered as components of a vector. Otherwise, you must do the plots one at a time by returning to the algebra window and separately highlighting each function.

Derive also does 2-dimensional polar plotting by changing the coordinate mode to Polar using the Options State command. In polar mode the expression is considered to be in the $r = f(\theta)$ or $f(\theta)$ form even though any variable names can be used.

Similarly, 2-dimensional parametric plotting is available by highlighting an expression in the form $[f(t), g(t)]$ with the x (horizontal) direction first and the y (vertical) direction second.

Derive's 3-dimensional rectangular surface plots are produced in a similar manner. The highlighted expression must be in the form $z = f(x,y)$ or just $f(x,y)$. Once again, the actual variable names are not important. When the Plot menu is selected, Derive recognizes the number of variables in the highlighted function and opens the 3-dimensional plot screen. The 3-dimensional plot menu is as shown in the following figure:

```
COMMAND: Algebra Center Eye Focal Grids Hide Length Options Plot Quit Window
         Zoom
Enter option
Center x:0           y:0           Length x:10      y:10      Derive 3D-plot
```

The commands to control the plot parameters for plotting surfaces in three dimensions are described in the following table.

Command	Function
Algebra	Sends control back to the algebra screen and menu.
Center	Positions the center of the plot at the specified coordinates.
Eye	Sets the coordinates of the viewer's eye.
Focal	Sets the coordinates of the focal point.
Grids	Establishes the number of grid panels in the x- and y-directions.
Hide	Allows for removal or inclusion of hidden lines.
Length	Establishes lengths of sides of the box where the plot resides.
Options	Allows control over axes display, colors of lines, and display mode.
Window	Opens a window. This is the same as the Window command in main menu.
Plot	Produces a plot of the highlighted function.
Zoom	Automatically changes the plot scale by a fixed amount. Zoom can be made In or Out.
Quit	Stops execution of the Derive program.

3-Dimensional Plot menu commands and their functions.

The coordinates of the plots center and length of sides are displayed in the status line at the bottom of the screen.

The following example shows a surface plot of the function $z(x,y)$ defined by
$$z(x,y) = \begin{cases} 5 - |x| - |y| & \text{if } 5 - |x| - |y| > 0 \\ 0 & \text{otherwise} . \end{cases}$$

This function can be converted to a single line expression in several ways using Derive functions. See the descriptions of the MAX and IF commands in Sections 1.6 and 1.11, respectively. Another way to define this function is with the STEP command in Derive. It is entered with the Author command and typing

$\boxed{\texttt{(5-abs(x)-abs(y)) STEP(5-abs(x)-abs(y))}}$.

The Plot menu is selected to open the 3-D plotting window. Then the default plot parameter Grid is changed to include 20 grid panels in both the x and y directions, the Length parameters are set to $x = 12$, $y = 12$, and $z = 5$, and the Eye parameters are set to $x = 20$, $y = 18$, and $z = 12$. Finally, the Plot command is issued to produce the following surface plot.

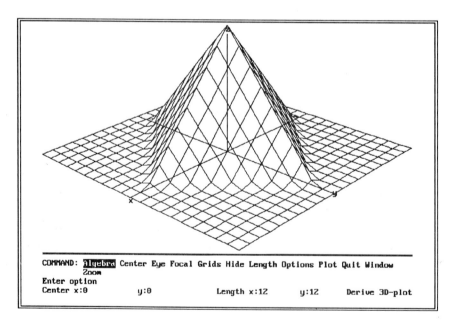

The utility file GRAPHICS.MTH contains functions that plot space curves and parametric surfaces. The actual plotting is done in the 2-D plot window. The AXES command builds the three dimensional axes in the 2-D plot window. Other commands in the file are ISOMETRIC, ISOMETRICS, SPHERE, TORUS, CYLINDER, CONE, RAYS, and ARCS. *Authoring* and

Simplifying expressions using these commands produce vectors that can be plotted in the 2-D rectangular plot mode with continuous lines. See the *Derive User Manual* for the specific functions and formats for these commands.

An example plot of the space curve

$$\vec{r}(t) = \sqrt{t}\cos(4t)\vec{i} + \sqrt{t}\sin(4t)\vec{j} + t/8\vec{k}, \quad 0 < t < 2\pi$$

which is entered as

$$\boxed{\texttt{ISOMETRIC([sqrt(t)cos(4t),sqrt(t)sin(4t),t/8])}}$$

is as follows:

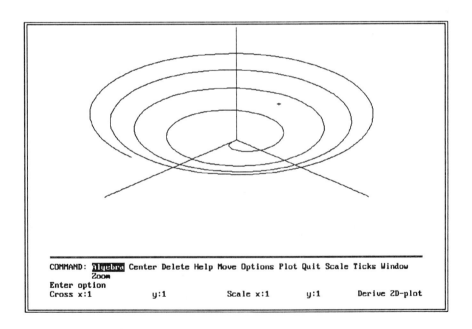

An example plot of a surface in the shape of a cone at an angle of $\pi/8$ radians from the *z*-axis, centered at the origin, and entered as

$$\boxed{\texttt{ISOMETRICS(CONE(pi/8,t,z),z,-2,2,4,t,-pi,pi,12)}}$$

is as follows:

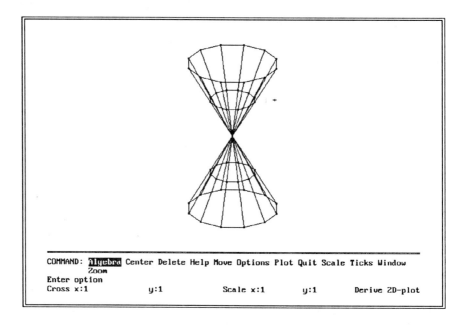

1.10 Numerical and Functional Approximations

We propose to use it as a "scientific exploration tool."

— John von Neumann [1946]

When Derive is unable to perform an exact computation or determine a solution symbolically, numerical approximations are viable alternatives. Derive has numerical capabilities in several areas. Some of these capabilities are discussed in this section.

If an antiderivative cannot be determined symbolically for the purpose of evaluating a definite integral, the user can change the Options Precision mode to Approximate to allow Derive to perform numerical quadrature. The numerical method is used automatically when Derive is in Mixed mode and an exact value cannot be calculated. The approximation method used is adaptive Simpson's quadrature. The method approximates the value of the definite integral with an error goal of the digital precision established with the Options Precision command. For example, Derive does not provide an answer to $\int_0^{0.2} e^x/(1-x^2)dx$ in Exact mode.

However, in Mixed mode with 6 digits of precision established, the following approximation is produced for this definite integral by using the Author command and entering

$$\boxed{\text{INT(Alt-e\textasciicircum x/(1-x\textasciicircum 2),x,0,0.2)}},$$

then executing Simplify.

$$\int_{0}^{0.2} \frac{e^{x}}{1-x^{2}} \, dx$$

$$0.224581$$

A similar circumstance can occur when solving for roots of an expression of one variable. When the equation is too difficult to solve completely in Exact mode, an implicit result is given. If the mode is changed to Mixed or Approximate, Derive uses the bisection method to find a root of the function to within the digital precision established. The user is prompted for the upper and lower bounds of the interval in which to search for the root. For example, if the roots of $e^x - x^4$ are required, in Exact mode no solution is produced. However, in either Mixed or Approximate mode the following result is obtained, if the bounds of 0 and 10 are provided.

$$17: \quad e^{x} - x^{4}$$

$$18: \quad x = 8.61314$$

The Calculus Taylor command finds a Taylor polynomial approximation to an expression. This command can be executed through the menu or in-line as TAYLOR($f(x), x, x_0, n$), where $f(x)$ is expanded in an nth degree polynomial about x_0. Taylor approximations can be handy for integration when direct symbolic integration is not possible.

It is usually best to use the Exact precision mode. When decimal representation of a rational or irrational number is needed, it can be obtained by using the approX command.

The utility file SOLVE.MTH has two functions to approximate roots of systems of equations. The function NEWTON(u,x,x0,n) iterates Newton's method n times to find the roots of the vector function $u(x)$. The vector x_0 contains the initial guess of the roots. The output is a matrix with $n+1$ rows, where each row is a successive iterate of the method. The approX command is used to execute this command.

The other approximation function in this file is FIXED_POINT(g,x,x0,n). This command produces n iterates of the vector equation $x = g(x)$ with the initial guess x_0. Like the NEWTON command, the resulting output is a matrix with $n+1$ rows of the successive iterates.

Derive is not designed to perform manipulation of large data sets. However, it has several functions that perform data analysis when the data is placed into a matrix with two or more columns. One such function is FIT(m) which provides a least squares approximation to the data in the functional form provided. See the *Derive User Manual* for a description of this command.

Another utility file that helps perform numerical operations is NUMERIC.MTH. This utility file contains commands to approximate first and second derivatives of functions with centered difference quotients and to approximate derivatives and integrals of given data arrays. The INT_DATA(A) command uses the trapezoidal rule to approximate the antiderivative of data arranged in the matrix A. Matrix A must contain 2 columns and each row of A represents an $[x,y]$ data point. See the *Derive User Manual* for descriptions of the other commands in this file.

1.11 Programming

We build systems [programs] like the Wright brothers built airplanes—build the whole thing, push it off a cliff, let it crash, and start over again.

— R. M. Graham [1970]

Derive is not a programming language. However, Derive has several commands that provide programming capabilities such as iteration and branching. These commands' formats and purposes are presented in the following table.

Command	Function
ITERATES(u,x,x0,n)	Computes the iterates of the recursive expression $x_{n+1} = u(x_n)$ for n steps starting with x_0. If the n is omitted, iteration continues until $x_{n+1} = x_n$. The approX command is used to execute this command.
ITERATE(u,x,x0,n)	Performs the same iteration as ITERATES, but returns only the last value of x_n.
IF(test,t,f,u)	*test* is a conditional test clause which is evaluated as true, false, or unknown. If *test* is true, the *t* expression is evaluated. If *test* is false, the *f* expression is evaluated. If *test* can not be determined (is unknown), then *u* is evaluated.

Commands that provide Derive with programming capabilities.

The IF command becomes more powerful through the use of nested IF expressions and the use of logical operators. The logical operators available are AND. OR, and NOT.

Simple examples of the use of these three commands are presented. If 10 iterates of $x_{n+1} = 2x_n + 6$, $x_0 = -2$, are needed, then Author

ITERATES(2x+6,x,-2,10).

Simplify to obtain:

```
20:   ITERATES (2 x + 6, x, -2, 10)
21:   [-2, 2, 10, 26, 58, 122, 250, 506, 1018, 2042, 4090]
```

If only the 8th iterate of the above expression is desired, Author

ITERATE(2x+6,x,-2,8).

Simplify to obtain:

```
22:    ITERATE (2 x + 6, x, -2, 8)

23:    1018
```

Given the conditional expression in two variables of

$$f(x,y) = \begin{cases} 0 & \text{if } x < 5 \text{ and } y \leq 2 \\ x & \text{if } x \geq 5 \text{ and } y > 3 \\ y & \text{otherwise} \end{cases}.$$

To enter this, Author

`f(x,y):=IF(x<5 AND y<=2,0,IF(x>=5 AND y>3,x,y),y).`

The resulting display is

```
24:    F (x, y) := IF (x < 5 AND y ≤ 2, 0, IF (x ≥ 5 AND y > 3, x, y), y)
```

Now in order to evaluate the expression for various values of x and y, the function f(x,y) is *Authored* and *Simplified*. For example, the input and output for the three pairs of x,y points (4,1), (6,4) and (2,4) is as follows:

```
25:    F (4, 1)

26:    0

27:    F (6, 4)

28:    6

29:    F (2, 4)

30:    4
```

The IF command can help in plotting discontinuous functions. For example, if the function of one-variable

$$g(x) = \begin{cases} x & \text{if } x < -1 \\ 2 & \text{if } -1 \leq x \leq 1 \\ -x & \text{if } x > 1 \end{cases}$$

is given, the command to Author the expression is

$$\boxed{\texttt{IF(x<-1,x,IF(x<=1,2,-x)).}}$$

This function can then be plotted by issuing Plot Plot. The following plot of this function was produced with an accuracy setting of 7:

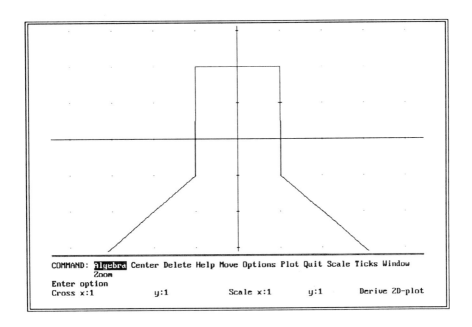

With this accuracy setting the discontinuities at $x = -1$ and $x = 1$ are not obvious, since the steep line connects the points on both sides of the discontinuity. By changing the accuracy setting to 9 with the Options Accuracy command and replotting, the following plot is produced. This new plot accurately shows the discontinuities in the function.

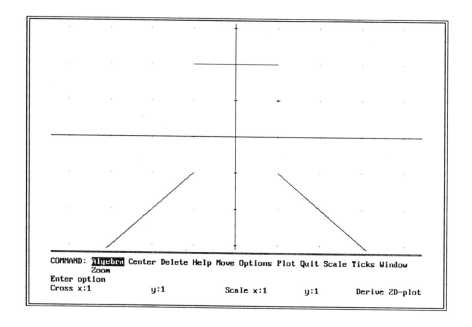

1.12 Demonstration of Commands and Capabilities

Computing offers a tool with which mathematics influences the modern world and a means of putting mathematical ideas into action.

— Lynn Arthur Steen [1990]

This section presents demonstrations of many of Derive's commands and capabilities. For the most part, the problems solved here are simple drill problems that can be solved directly by a single Derive command. The usual method used for these problems is to Author an expression and then Simplify the expression. The format for each demonstration problem is as follows: 1) problem statement, 2) Derive input (in-line commands are outlined in a box), 3) the output display from Derive that shows the solution of the problem, and 4) information and comments about the problem, if needed.

Demonstration of Commands and Capabilities

Demonstration #1. Expand $(x+y)^4$. Author $\boxed{\text{(x+y)^4}}$. Press **Enter**. Select Expand and press **Enter** (blank) for the Expand Variable.

```
              4
1:     (x + y)

       4     3       2 2       3     4
2:    x + 4 x y + 6 x y  + 4 x y  + y
```

Demonstration #2. Factor $x^3 + 2x^2 - 2x + 3$.
Author $\boxed{\text{x^3+2x^2-2x+3}}$. Press **Enter**. Select Factor. Then select Complex.

$$3:\quad x^3 + 2x^2 - 2x + 3$$

$$4:\quad \left[x - \frac{1}{2} - \frac{\sqrt{3}\,\hat{\imath}}{2}\right]\left[x - \frac{1}{2} + \frac{\sqrt{3}\,\hat{\imath}}{2}\right](x+3)$$

Demonstration #3. Find the prime factors of 5858743749.
Author $\boxed{5858743749}$. Press **Enter**. Select Factor.

```
5:    5858743749

6:    3  41  1553  30671
```

Demonstration #4. Evaluate and approximate $\ln(\sqrt{2})$.
Author $\boxed{\text{ln(sqrt(2))}}$. Press **Enter**. Select Simplify.

$$7:\quad \mathrm{LN}(\sqrt{2})$$

$$8:\quad \frac{\mathrm{LN}(2)}{2}$$

Execute approX.

42 Chapter 1. Getting Started

```
7:      LN (√2)

        LN (2)
8:      ──────
          2

9:      0.346573
```

The format of the output depends upon the settings in the `Options-Notation` submenu. The options set of this output are highlighted on the following menu display.

```
NOTATION: Style: Decimal Mixed Rational Scientific  Digits: 6

Enter numerical output style
Approx(8)                                           Free:96%
```

Demonstration #5. Find $\lim_{x \to 0} \sin(x)/x$.
Author `LIM(sinx/x,x,0)`. Press **Enter**. Select `Simplify`.

```
                    SIN (x)
10:     lim         ───────
        x→0            x

11:     1
```

Another method to evaluate limits can be performed using the `Calculus` menu command. See Sections 1.7 and 2.1.

Demonstration #6. Find the first derivative of $x \sin(3x^3 - x)$.
Author `DIF(x sin(3x^3-x),x)`. Press **Enter**. Select `Simplify`.

```
         d              3
12:     ── x SIN (3 x  - x)
        dx

             2             3             3
13:   x (9 x  - 1) COS (3 x  - x) + SIN (3 x  - x)
```

Another method to evaluate derivatives can be performed using the Calculus menu command. See Sections 1.7 and 2.3.

Demonstration #7. Find the second derivative with respect to x of $|x|\cos x - e^{ax}$.

Author `DIF(abs(x)cosx-Alt-e^(ax),x,2)`.

Press **Enter**. Select `Simplify`.

$$14: \left[\frac{d}{dx}\right]^2 (|x| \cos(x) - e^{ax})$$

$$15: -\text{SIGN}(x)(x\cos(x) + 2\sin(x)) - a^2 e^{ax}$$

Demonstration #8. Find $\int x\cos^2 x\, dx$.

Author `INT(xcos^2(x),x)`. Press **Enter**. Select `Simplify`.

$$16: \int x \cos(x)^2 \, dx$$

$$17: \frac{x \sin(x) \cos(x)}{2} - \frac{\sin(x)^2}{4} + \frac{x^2}{4}$$

Another method to evaluate integrals can be performed using the Calculus menu command. See Sections 1.7 and 2.6.

Demonstration #9. Find $\int_1^\infty \frac{1}{x^2}$.

Author `INT(1/x^2,x,1,inf)`. Press **Enter**. Select `Simplify`.

$$18: \int_1^\infty \frac{1}{x^2}\, dx$$

$$19: 1$$

Demonstration #10. Find $\sum_{k=1}^{10} 5k^3$.

Author `SUM(5k^3,k,1,10)`. Press **Enter**. Select `Simplify`.

```
              10   3
       20:    Σ   5 k
              k=1

       21:    15125
```

Another method to evaluate summations can be performed using the Calculus menu command. See Sections 1.7 and 2.20.

Demonstration #11. Find the 5-term Taylor polynomial approximation to $e^{-x}\cos(x^2)$ about the point $x = 0$.

Author TAYLOR(Alt-e^-x cos(x^2),x,0,5).

Press **Enter**. Select Simplify.

```
                    -x      2
       22:  TAYLOR (ê   COS (x ), x, 0, 5)

              5        4      3     2
            59 x     11 x    x     x
       23:  ───── -  ───── - ─── + ─── - x + 1
             120      24     6     2
```

Another method to determine Taylor polynomials can be performed using the Calculus menu command. See Sections 1.7 and 2.13.

Demonstration #12. Evaluate the function $\sqrt{x}e^{2x}\csc(\pi x)$ for $x = 4.5$.

Author sqrt(x)Alt-e^(2x)csc(pi x). Press **Enter**. Execute Manage Substitute. Press **Enter**. Replace x with 4.5. Select Simplify.

```
                      2 x
       24:    √x ê        CSC (π x)

                        2 4.5
       25:    √4.5 ê         CSC (π 4.5)

                     9
              3 √2 ê
       26:   ────────
                 2
```

Demonstration #13. Evaluate $2^{32} - 1$ and factor it into prime factors. Author $\boxed{2\char`\^32-1}$. Press **Enter**. Select **Expand**. Select **Factor**.

```
27:   2^32 - 1

28:   4294967295

29:   3 5 17 257 65537
```

1.13 Utility Files for Calculus

As is well known, physics became a science only after the invention of differential calculus.

— Bernhard Riemann [1882]

In this section, we will discuss the commands found in three utility files: DIF_APPS.MTH, INT_APPS.MTH, MISC.MTH. These functions give Derive added capabilities in performing calculus operations. The help file contains brief descriptions the functions. Most of the functions in these three files and their purposes are given in the following tables. More information about these commands are found in the *Derive User Manual*. Many of these commands are used in the examples in Chapter 2.

Function Name	Purpose
CURVATURE(y,x)	Finds the curvature formula for the equation $y = f(x)$.
CENTER_OF_CURVATURE(y,x)	Finds the vector formula for the location of the center of curvature.
TANGENT(y,x,x0)	Finds the equation of the line tangent to $y = f(x)$ at $x = x_0$.
PERPENDICULAR(y,x,x0)	Finds the equation of the line perpendicular to $y = f(x)$ at $x = x_0$.
OSCULATING_CIRCLE(y,x,t)	Finds the parametric representation of the circle that osculates $y = f(x)$ at x.

Basic functions for applying the derivative to find properties of expressions in Cartesian coordinate form found in the file DIF_APPS.MTH.

Function Name	Purpose
PARA_DIF(v,t,n)	Finds $d^n y/dx^n$ in terms of the parametric variable t. v is a vector expression $[x(t), y(t)]$.
PARA_CURVATURE(v,t)	Finds the formula for the curvature of v in terms of t.
PARA_CENTER_OF_CURVATURE(v,t)	Finds the vector expression for the location of the center of curvature in terms of t.
PARA_TANGENT(v,t,t0,x)	Finds the equation in x of the line tangent to the curve $v(t)$ at $t = t_0$.
PARA_PERPENDICULAR(v,t,t0,x)	Finds the equation in x of the line perpendicular to the curve $v(t)$ at $t = t_0$.
PARA_OSCULATING_CIRCLE(v,t,t0,θ)	Finds the parametric representation of the circle that osculates $[x, y] = v$ at t_0.

Basic functions for applying the derivative to find properties of expressions in parametric form found in the file DIF_APPS.MTH.

Function Name	Purpose
POLAR_DIF(r,θ,n)	Finds $d^n y/dx^n$ in terms of the polar variable θ. r is a function of θ.
POLAR_CURVATURE(r,θ)	Finds the formula for the curvature.
POLAR_CENTER_OF_CURVATURE (r,θ)	Finds a vector expression for the location of the center of curvature.
POLAR_TANGENT(r,θ,θ_0,x)	Finds the equation in x of the line tangent to $r(\theta)$ at $\theta = \theta_0$.
POLAR_PERPENDICULAR (r,θ,θ_0,x)	Finds the equation in x of the line perpendicular to $r(\theta)$ at $\theta = \theta_0$.
POLAR_OSCULATING_CIRCLE (r,θ,$\theta 0$,ϕ)	Finds the parametric representation of the circle that osculates the polar curve r at $\theta = \theta_0$.

Basic functions for applying the derivative in polar coordinates from file DIF_APPS.MTH.

DIF_APPS contains two commands that find special properties of 3-dimensional surfaces defined implicitly by $u(x, y, z) = 0$. The following table explains the purpose and form of these two commands which are applications of the derivative.

Function Name	Purpose
TANGENT_PLANE(u,v,v0)	Finds the equation of the plane tangent to $u = 0$ at $v = v_0$. $v = [x, y, z]$.
NORMAL_LINE(u,v,v0,t)	Finds the parametric form of the vector that defines the line perpendicular to $u = 0$ at $v = v_0$.

Functions for finding two properties of a surface from utility file DIF_APPS.MTH.

DIF_APPS also contains several analogous commands to those commands in the previous 3 tables for implicit functions defined in Cartesian coordinates ($u(x, y) = 0$). No table is given here for these commands. See the *Derive User Manual* for their description.

The commands that apply or use the integral are located in utility file INT_APPS.MTH. These commands are discussed in the following table. When the integrals cannot be evaluated exactly using the Simplify command, try a numerical approximation method by selecting the approX command.

Function Name	Purpose
ARC_LENGTH(y,x,x1,x2,u)	Finds the arc length of a curve $y(x)$ with x varying from x_1 to x_2.
POLAR_ARC_LENGTH($r,\theta,\theta_1,\theta_2$,u)	Finds the arc length of a polar curve $r(\theta)$ with θ varying from θ_1 to θ_2.
PARA_ARC_LENGTH(v,t,t1,t2,u)	Finds the arc length of a parametric curve $v(t)$ with t varying from t_1 to t_2.
AREA(x,x1,x2,y,y1,y2,u)	Finds the area between the functions $y = y_1(x)$ and $y = y_2(x)$ with x varying from x_1 to x_2.
AREA_CENTROID (x,x1,x2,y,y1,y2,u)	Computes the centroid of the region described above.
POLAR_AREA(r,r1,r2,θ,θ_1,θ_2,u)	Finds the area between the polar functions $r = r_1(\theta)$ and $r = r_2(\theta)$ with $\theta_1 \leq \theta \leq \theta_2$.
SURFACE_AREA (z,x,x1,x2,y,y1,y2,u)	Finds the surface area or surface integral defined by $z = f(x,y)$
VOLUME (x,x1,x2,y,y1,y2,z,z1,z2,u)	Finds the volume or volume integral (mass) of a region.
VOLUME_CENTROID (x,x1,x2,y,y1,y2,z,z1,z2,u)	Finds the centroid of a body with density u.
SPHERICAL_VOLUME (r,r1,r2,θ,$\theta1,\theta2,\phi,\phi1,\phi2$,u)	Finds the volume integral of a shape defined in spherical coordinates.
CYLINDRICAL_VOLUME (r,r1,r2,$\theta,\theta1,\theta2$,z,z1,z2,u)	Finds the volume integral of a shape defined in cylindrical coordinates.

Basic functions for applying the integral found in the utility file INT_APPS.MTH. The u is an optional density function for most of the commands.

INT_APPS also contains functions for computing the moments of inertia and for the advanced techniques of Laplace transform and Fourier series. See the *Derive User Manual* for information about these two functions.

The calculus commands in MISC.MTH are discussed in the following table. This file also contains commands that assist with performing mathematical (inductive) proofs. See the *Derive User Manual* for descriptions of those commands.

Function Name	Purpose
MOD(a,b)	Computes a **modulo** b or the remainder when a/b.
FLOOR(a,b)	Finds the greatest integer less than or equal to a/b.
RATIO_TEST(t,n)	Computes a value u. If $u > 1$, $\sum_{n=a}^{\infty} t(n)$ diverges. If $u < 1$, the sum converges. $t(n)$ must be positive.
LIM2(u,x,y,x0,y0)	Evaluates the 2-dimensional limit of $u(x,y)$ as $[x,y] \to [x_0, y_0]$ along a line of slope @1. If the result does not contain @1, the limit is the same for all slopes.
LEFT_RIEMANN(u,x,a,b,n)	Computes the left Riemann sum of $u(x)$ from $x = a$ to $x = b$ using n equally spaced intervals.
INT_PARTS(u,dv,x)	Evaluates the integration by parts formula $\int uv = u \int (dv) dx - \int (\int (dv) dx) \frac{du}{dx} dx$.
INT_SUB(y,x,u)	Performs the integration of y by u-substitution.
INVERSE(u,x)	Finds $u^{-1}(x)$.

Basic functions for applying the integral found in file MISC.MTH.

1.14 Differential Equations

A mathematician, like a painter or poet, is a maker of patterns. If his patterns are more permanent than theirs, it is because they are made with ideas.
— Godfrey Hardy [1940]

There are three utility files that contain functions to help solve differential equations. File ODE1.MTH is for first-order equations, and ODE2.MTH contains commands for second-order equations. File ODE_APPR.MTH contains numerical approximation methods for differential equations. Several of the functions in these files are used in examples in Chapter 2. The book, *Derive Laboratory Manual for Differential Equations*,[24] has numerous examples and exercises of using Derive to solve differential equations. The proper type and form of the equation must be determined before the functions can be used. These built-in functions make Derive a powerful and efficient tool for solving differential equations. Some of the functions in these three files and their purposes are given in the following tables. Further details are provided for functions when they are used in examples.

Function Name	Purpose
SEPARABLE(p,q,x,y,x0,y0)	Solves a separable differential equation in the form $y' = p(x)q(y)$. Solution is in an implicit form.
EXACT_TEST(p,q,x,y)	Checks if $p(x,y) + q(x,y)y' = 0$ is exact. If the result is 0, the equation is exact.
EXACT(p,q,x,y,x0,y0)	Solves an exact equation (above), with $y(x0) = y0$. Solution is in implicit form.
USE_INTEG_FCTR (m,p,q,x,y,x0,y0)	Solves an equation with known integrating factor m.
LINEAR1(p,q,x,y,x0,y0)	Solves a linear equation of the form $y' + p(x)y = q(x)$ with $y(x_0) = y_0$.

Basic functions for solving first-order equations found in the utility File ODE1.MTH.

---

[24]Written by the same author as this book and published by Addison-Wesley in 1991.

Function Name	Purpose and/or Form of Equation to be Solved
LIN2_TEST(p,q,r)	Finds the value of discriminant k, which determines which of the next 3 commands should be used to solve $y'' + p(x)y' + q(x)y = r(x)$.
LIN2_POS(p,q,r,x)	Solves $y'' + p(x)y' + q(x)y = r(x)$, if above $k > 0$.
LIN2_NEG(p,q,r,x)	Solves $y'' + p(x)y' + q(x)y = r(x)$, if above $k < 0$.
LIN2_0(p,q,r,x)	Solves $y'' + p(x)y' + q(x)y = r(x)$, if above $k = 0$.
IMPOSE_BV2 (x,y,x0,y0,x2,y2,c1,c2)	Given $y(x, c_1, c_2), y(x_1) = y_1, y(x_2) = y_2$, finds c_1 and c_2.
IMPOSE_IV2 (x,y,x0,y0,v0,c1,c2)	Given $y(x, c_1, c_2), y(x_0) = y_0, y'(x_0) = v_0$, finds c_1 and c_2.

Functions and the form of applicable second-order equations which can be solved using some methods found in utility File ODE2.MTH.

Utility files ODE1 and ODE2 contain several more functions to solve some equations with special forms and more complicated equations. The following table gives information about the approximation methods found in file ODE_APPR.MTH.

Function Name	Purpose and/or Form of Equation to be Solved
TAY_ODE1(r,x,y,x0,y0,n)	Finds the nth degree Taylor-series solution to $y' = r(x,y), y(x_0) = y_0$.
PICARD(r,p,x,y,x0,y0)	Given an approximate solution p to $y' = r(x,y), y(x_0) = y_0$, and finds an improved iterate.
EULER(r,x,y,x0,y0,h,n)	Uses Euler's method to approximate the solution to $y' = r(x,y), y(x_0) = y_0$, at n values of x, $(x_0, x_0 + h, ..., x_0 + nh)$. h is the step size of the calculations.
TAY_ODES(r,x,y,x0,y0,n)	Finds the nth degree Taylor-series solution to a system of differential equations.
RK(r,v,v0,h,n)	Uses the Runge-Kutta method to approximate the solution to a first-order equation $(y' = r(x,y))$ or system of such equations for n steps with a step size of h.
DIRECTION_FIELD (r,x,x0,xm,m,y,y0,yn,n)	Produces a matrix of vectors that plots a direction field for $y' = r(x,y)$. There are m vectors in the x-direction, and n vectors in the y-direction.

Functions that solve first-order equations in utility File ODE_APPR.MTH.

In order to use any of these functions, the appropriate utility file (ODE1, ODE2, or ODE_APPR) must be loaded into the work area using either the Transfer Load Utility or Transfer Merge commands. Users can make their own utility files for solving differential equations by keeping the most useful commands in the work area and saving them for use in the future using the Transfer Save command.

1.15 Limitations

For it is unworthy of excellent men to lose hours like slaves in the labor of computation.

— Gottfried Wilhelm Leibniz

All software packages have limitations, and Derive is no exception. It is usually best to be warned of those limitations before discovering them first-hand at a critical step in solving a problem.

Probably Derive's most obvious restriction is its limited programming capability. Derive's commands are executed one at a time with user interface needed after each step. There is no way to execute loops of multi-line commands. However, through the use of nested functional calls, several operations can be executed at once. Despite this limitation, there are ways to do iteration (ITERATES command), recursion, and branching (IF command). See Section 1.11 for a discussion of these commands. An inconvenience related to this situation is that subscripted variables use a rather awkward ELEMENT function to access elements.

Because the software is mostly menu-driven, the user interface is crucial. Yet Derive does not support the use of a mouse to facilitate this interface. This sometimes slows cursor movement, highlighting, and selection of menu options.

Derive's work area is designed for executable commands, not text. Therefore, there is very little text formatting allowed in the work area. This restricts the quality of comments in the screen image and printed output.

The limitations in Derive's plotting tools are the *lack* of i) direct entry of domain bounds for a rectangular plot, ii) an automatic scaling option for two-dimensional plots, iii) the ability to label axes and designate the axes to specific variables, iv) the ability to handle implicit functions, and v) the ability to plot level curves for a surface.

The only task that sometimes seems slow in Derive is plotting. Despite the user's ability to change accuracy parameters to speed up plotting, it sometimes seems that the plots take too long. This is especially true if several plots are needed. Although, the ability to plot a vector of expressions with one plot command is very helpful.

1.16 Recommendations

Old people like to give good advice, as solace for no longer being able to provide bad examples.

— François, Duc de La Rochefoucald [1678]

There are several things that can be done to insure efficient and proper use of the tremendous capabilities of Derive and to tailor the software to the specific needs of the user. First, insure the system state you use most often is saved into the DERIVE.INI file and, therefore, is in effect when the program is started. This is accomplished by putting the computer in the state you want and then saving that state using the `Transfer Save State` command. If you use the plotting features often, insure this initial state is in the appropriate `Graphics` mode. This eliminates the need to change states very often in order to get the capabilities you need. See Sections 1.2 and 1.5 and the *Derive User Manual* for more information on this feature.

Next, build and save utility files that contain the commands that you use in a manner that is efficient for your use. For instance, select the commands you use most often in the utility files DIF_APPS.MTH, INT_APPS.MTH, ODE1.MTH, ODE2.MTH, and ODE_APPR.MTH and design your own utility file for solving calculus problems. You may want to add other functions you frequently use from MISC.MTH or SOLVE.MTH to your personal utility file. You may even want to build several special utility files. For example, you may want one for operations in Cartesian coordinates, a second for operations on parametric functions, and a third for polar functions. The following table lists some functions, along with the Derive utility file they are found in, that could produce a nice utility file for a calculus course.

Function Name	Utility File
CURVATURE	DIF_APPLS
TANGENT	DIF_APPLS
PERPENDICULAR	DIF_APPLS
ARC_LENGTH	INT_APPLS
AREA	INT_APPLS
SEPARABLE	ODE1
EXACT_TEST	ODE1
EXACT	ODE1
LINEAR1	ODE1
LIN_TEST	ODE2
LIN_NEG	ODE2
LIN_0	ODE2
LIN_POS	ODE2
EULER	ODE_APPR
RK	ODE_APPR

Suggested functions to include in a utility file for a calculus course.

If you frequently repeat a sequence of commands or use a command in a special way, try to design your own commands and add them to appropriate utility files.

Finally, spend some time learning the structure of the menus and the keystrokes that help manipulate expressions and reduce retyping expressions. The menu structure is discussed in Section 1.2. The keystrokes are discussed in Section 1.3. Derive's menu system makes it user-friendly; however, the understanding and use of a few of the special keystrokes can turn you into a more efficient user of this powerful software tool.

Probably the most overlooked and underutilized features are:
1) Using **F3**, **F4**, and expression labels (such as #21) in the author line.
2) Using **Ctrl-S** and **Ctrl-D** to move the cursor within the author line.
3) Highlighting a subexpression (part of an existing expression) before using `Simplify`, `Expand`, `Factor`, `approX`, or `Manage Substitute`.
4) Using the `approX` command instead of `Simplify` when desiring a numerical approximation.

2
Examples

The heart of mathematics consists of concrete examples and concrete problems.

—Paul R. Halmos [1970]

These examples have been worked out to show the power and versatility of Derive as a problem-solving tool. By carefully reading these problems and solutions and working along with Derive, the reader should get a better feel for the subject of calculus and of mathematics in general. Many of these examples involve realistic applications, while others are posed in a mathematical context. These problems are similar to those typically found in calculus textbooks. References to the sections in Chapter 1 that contain helpful information are provided to expedite the learning of Derive's capabilities and to provide efficiency in the use of this manual.

Example 2.1 Limits and Continuity

There are things which seem incredible to most men who have not studied mathematics.

—Archimedes [c. 287-212 B.C.]

Subject: Finding the local behavior, continuity, and limits of functions

58   Chapter 2. Examples

References: Sections 1.7 and 1.9

Problem: Explore the behavior of the following functions near the indicated values:

$$f(x) = \frac{x}{|x|} \quad \text{at} \quad x = 0$$

$$g(x) = \frac{x^2 - 3x - 4}{x^2 - 1} \quad \text{at} \quad x = 1 \text{ and } x = -1$$

$$h(x) = \frac{\cos(x) - 1}{x} \quad \text{at} \quad x = 0$$

$$p(x, y) = \frac{xy}{x^2 + y^2} \quad \text{at} \quad (0, 0) \quad \text{along any line.}$$

Solution: A graph is often worth more than a thousand numbers or a thousand calculations and gives us geometrical intuition about the problem. So our first step is to graph the functions near the x values of interest. To do this for $f(x)$, select the Author command and type x/abs(x). This displays in the work area as:

$$1: \quad \frac{x}{|x|}$$

To obtain a plot of this function, select the Plot submenu. Execute the Plot command in this menu to obtain a graph of the highlighted function using the default plotting parameters. Depending on the values of those parameters, the graph is as follows:

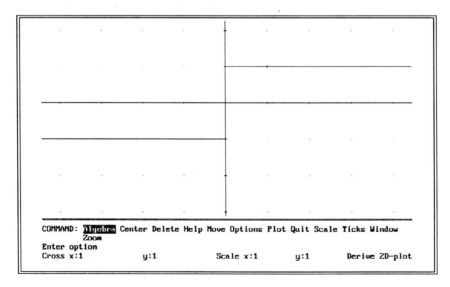

The plot indicates there is a jump discontinuity in $f(x)$ at $x = 0$. To be sure, we can take the limit of $f(x)$ as x approaches 0 from both the left and right sides. First, return control back to the Algebra window. Then, use the Calculus Limit command, and select $x = 0$ and the Left direction. Remember the Tab key advances control to the next part of the menu. Select Simplify to get the $\lim_{x \to 0^-} f(x)$. The result is

```
1:      x
       ───
       |x|

2:  lim   x
    x→0-  ───
          |x|

3:  -1
```

In order to obtain the right limit, follow a similar procedure. First, use the direction arrow key ↑ to highlight the function in expression #1. Then, execute the Calculus Limit command, and select $x = 0$ and the Right direction. Select Simplify to get the $\lim_{x \to 0^+} f(x)$. This time the result is

```
4:  lim   x
    x→0+  ───
          |x|

5:  1
```

These different limit results verify the jump discontinuity.

Let's follow a similar procedure for $g(x)$. Author $\boxed{\text{(x^2-3x-4)/(x^2-1)}}$.

The display shows:

```
         2
        x  - 3 x - 4
6:      ─────────────
             2
            x  - 1
```

Execute `Plot Delete All` to get rid of the previous plots. The execute `Plot` to obtain the new plot of $g(x)$ as follows:

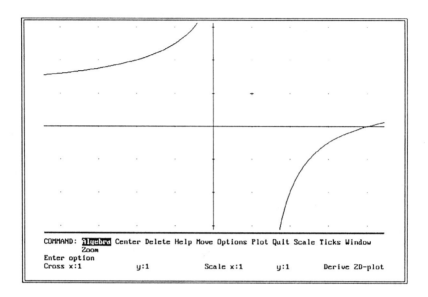

The behaviors of $g(x)$ near $x = 1$ and $x = -1$ appear quite different; yet, both of these values of x produce a 0 in the denominator of $g(x)$. Let's explore the behavior by taking limits. Execute `Calculus-Limit` and select the point $x = -1$ and Both directions. Select `Simplify`. The result is:

$$7: \quad \lim_{x \to -1} \frac{x^2 - 3x - 4}{x^2 - 1}$$

$$8: \quad \frac{5}{2}$$

The function is discontinuous at $x = -1$.

For $x = 1$, we use the other set-up procedure involving the in-line LIM command. Author

`LIM((x^2-3x-4)/(x^2-1),x,1,0)`.

The last parameter setting of 0 produces the limit from both sides. Simplify to perform the limit operation. The command and its result are:

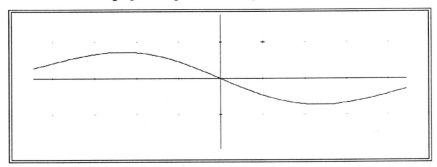

This time we see that $g(x)$ "blows up" to infinity (indicated by 1/0) at $x = 1$.

Now, let's explore $h(x)$. First, Author $\boxed{(\cos x\ -1)/x}$. Then execute Plot Plot. The graph using the default parameters is as follows:

Since the x value of interest is 0, let's zoom in on the function at $x = 0$. Do this by moving the cross to (0,1) by selecting the Move command and setting $x = 0$ and $y = 1$. Then execute the Center command. Select Zoom and set the parameters to In and Both. Remember that the **Tab** key toggles between the different parts of the menu. After zooming in the plot looks as follows:

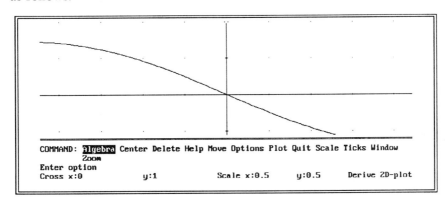

This seems to show that the $\lim_{x \to 0} h(x) = 0$, yet the denominator is 0 for $x = 0$. Return to the Algebra window. Set up the limit as $x \to 0$ from both sides by the Calculus Limit menu command or the LIM in-line command. Simplify to confirm what the graph suggested as shown in the following Derive display:

```
12:  lim   COS (x) - 1
     x→0   ──────────
               x

13:  0
```

Finally, we investigate the two-dimensional function $p(x, y) = \frac{xy}{x^2+y^2}$. In order to take a limit in two variables the utility file MISC.MTH must be loaded. Execute the Transfer Load Utility command and enter MISC for the filename. Author the limit command

$\boxed{\text{LIM2(xy/(x^2+y^2),x,y,0,0)}}$.

Simplify to get the following result:

```
              ⎡   x y              ⎤
16:  LIM2    ⎢ ───────, x, y, 0, 0 ⎥
              ⎣  2    2            ⎦
                 x  + y

          @1
17:    ───────
         2
        @1 + 1
```

The solution contains 1 which means the limit depends on the slope (1) of the lines through (0,0). Therefore, the limit of the function at the point does not exist.

Example 2.2 Curve Sketching

A picture shows me at a glance what it takes dozens of pages of a book to expound.

— Ivan Turgenev [1862]

Subject: Curve sketching and analyzing function behavior
References: Sections 1.9 and 1.13
Problem: Sketch the following functions and categorize their local and global behavior:

$$f(x) = \frac{1}{x} \sin(x^2)$$

$$g(x) = \sin x + x$$

$$h(x) = \tanh(1 - 20x)$$

$$s(x) = \text{Floor}(x^2, 1) = \lfloor x^2 \rfloor = \text{integer part of } x^2$$

Solution: Often the most important step in solving a problem is sketching the curve or function. Derive and Calculus complement each other as aids to do curve sketching, especially as it relates to finding the most important features of a curve — local maximums, local minimums, intercepts, and asymptotes. Producing plots in Derive is easy. The challenge is to produce a plot of the function with the correct scale and viewing window to gain good geometrical intuition about the behavior of the function.

The first step to plot a function is to Author the function into the work area. The entry of $f(x)$ is $\boxed{\texttt{1/x sin(x\^{}2)}}$. Then select the Plot submenu. The actual plot region you see in the display depends upon the plotting parameters set with the commands in the Plot submenu. Before we plot $f(x)$, let's reset the plot parameters to try to capture the global behavior of $f(x)$. First, we will center the plotting region at (0,0). Do this by executing Move and setting $x = 0$ and $y = 0$. Remember that the **Tab** key provides for the movement between parts of the menu. Then select Center to make (0,0) the center of the plotting region. In order to obtain a global perspective by seeing more of the change in the x-direction, the x-axis scale must be increased to a higher value than the y-axis scale. Execute the Scale command and set the x-axis scale to 3 and the y-axis scale to 1. Now select Plot to acquire the following graph:

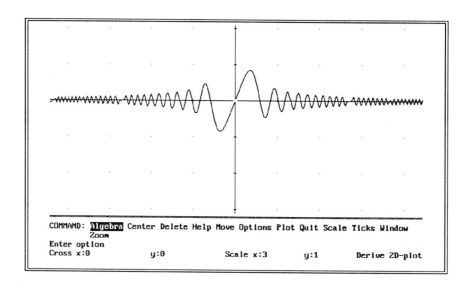

This function seems to oscillate with decaying amplitude the further it gets from the origin. We also notice some strange behavior (oscillations are interrupted slightly) at a couple places. Further investigation, by zooming in and out at those locations, shows this to be aliasing caused by the graphing procedure and the scale being used. Aliasing is the appearance of a behavior on the graph that is not actually present in the function. It is caused by interpolation in the graphing procedure along with the large scale being used.

To get a plot of the local behavior near a point of interest, we need to change the scale and move the center of the plot. Rescale as we did before. This time set x-scale $= 0.25$ and y-scale $= 0.25$. Let's see the behavior near $x = 10$, so select the Move command and set $x = 10$ and $y = 0$. Issue the Center command. The plot is redrawn with these plotting parameters. At this interval and scale, the function locally looks like a normal sine curve with a reduced amplitude.

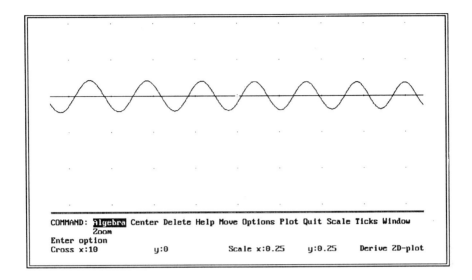

Execute Delete All to clear the plot window. Select Algebra to return control to the algebra window.

Let's attack $g(x)$ in a similar fashion. Author $\boxed{\text{sinx+x}}$. Issue Plot to obtain the 2-dimensional plotting window. For a global view of the function, select Move to move the cross to $x = 0, y = 0$ and execute Center. Set the scale using the Scale command to $x = 10$ and $y = 10$ per tick mark on the axes. This provides a plotting region of approximately $-40 < x < 40$ and $-30 < y < 30$. Issue the Plot command to obtain the following graph of $y = g(x)$.

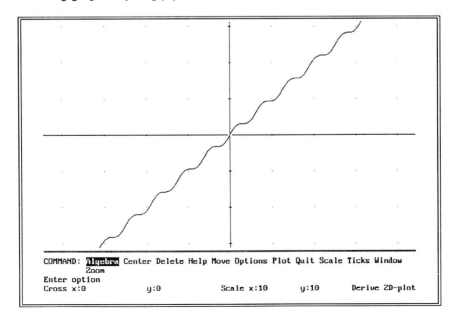

66   Chapter 2. Examples

The graph of $g(x)$ globally resembles the straight line $y = x$. Larger scales will eliminate the wiggles altogether.

To investigate the local behavior near (0,0), we must rescale the tick marks. Select the `Scale` submenu and set $x = 1.5$ and $y = 1.5$. The **Tab** key moves control through different parts of the submenu. The plot is automatically redrawn with the new scale when the **Enter** key is typed. The plot showing the local behavior is as follows:

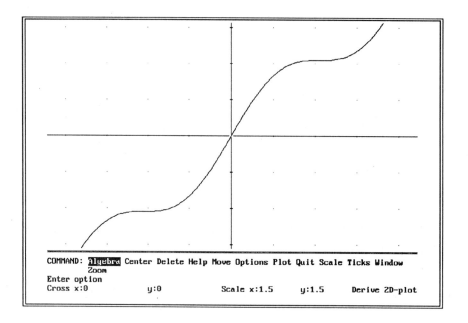

For the function $h(x)$ we will do something a little different. We will split the screen into two windows: one to show the sketch of $h(x)$ and the second window to display the functional definition. First Author `tanh(1-20x)`. Then select `Window Split Horizontal`. Then answer the query to split the screen at line 5. Notice that the two windows are numbered in the upper-left corners. The top window (#1) has its number highlighted which indicates that window is the active one. The **F1** key toggles control between the two windows. (See Sections 1.4 and 1.9 for more information about using windows and performing plotting.)

Press **F1** to activate window #2. Now we will designate this window for plotting. We do this through the `Window Designate 2D-Plot` command. Answer Y for yes to the query to abandon the current algebraic expressions in this window. The algebraic expressions will remain in window #1. The screen is now split with part of the screen set up for plotting and appears as follows:

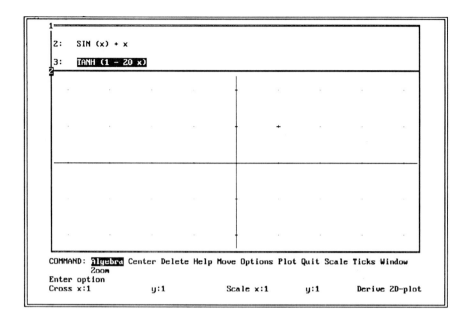

Now we can perform the graphing of $h(x)$ by executing Plot. The resulting graph is shown.

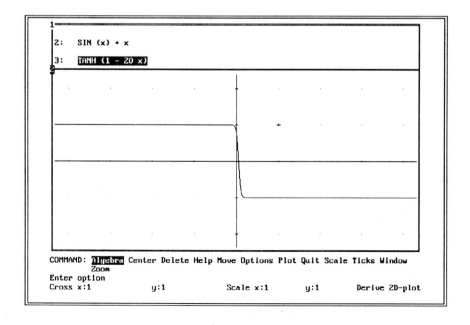

Globally this function resembles a step function. It appears to be 1 for $x < 0$ and -1 for $x > 0$ with a very steep, near-vertical line connecting these values near or at $x = 1$. Before we jump to any conclusions, let's get a closer look near $x = 0$. Select Scale and set the x-scale to 0.05. Press Enter to obtain this new, smaller-scaled plot.

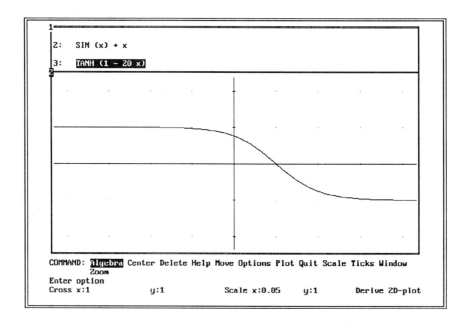

On this finer scaled plot, we see that the transition from 1 to -1 takes place in very smoothly and the x-intercept is not 0, but a value close to 0.05. Actually, $h(x)$ is locally quite different from a discontinuous step function.

Now we are ready to tackle $s(x)$. The FLOOR function is in the utility file MISC.MTH, so we must Transfer Load Utility the commands from that utility file into the work area. Then Author

FLOOR(x^2,1).

Then execute Plot. Select Scale and set x-scale to 1 and y-scale to 5. The Options Accuracy must be set to 9 to get a pointwise accurate plot. Then execute Plot to get:

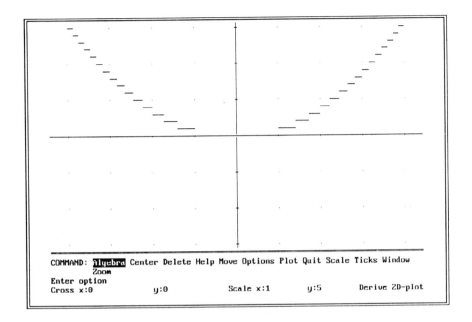

Try other settings for the accuracy parameter to see the problems that can occur in plotting discontinuous functions.

Example 2.3 Derivatives

But the velocities of the velocities — the second, third, fourth, and fifth velocities, etc. — exceed, if I mistake not, all human understanding.

—George Berkeley [1734]

Subject: Using derivatives to analyze functions
Reference: Section 1.7
Problem: Use the appropriate derivatives and analysis to find the points of local and global maximum and minimum and the points of inflection for the following functions:

$$f(x) = x^3 - 4x^2 - 6x - 1/2$$

$$g(x) = \frac{x^2 + 1}{x^2 - 4x + 3}$$

$$h(x) = x \ln x + \sin x$$

Solution: Finding the derivative of functions using Derive is simple and direct. There are two ways to obtain a derivative of a function. See Section 1.7 for more information. The first way is to Author the function into the work area. Then select the Calculus menu. Select Differentiate and input the parameters for order and variable of differentiation. The keystrokes for finding $\frac{df}{dx}$ are as follows: Author $\boxed{\text{x}^\wedge\text{3-4x}^\wedge\text{2-6x-1/2}}$.

Execute Calculus Differentiate and respond with variable of differentiation x and order of 1. Then execute Simplify. The resulting output is as follows:

$$1: \quad x^3 - 4x^2 - 6x - \frac{1}{2}$$

$$2: \quad \frac{d}{dx}\left[x^3 - 4x^2 - 6x - \frac{1}{2}\right]$$

$$3: \quad 3x^2 - 8x - 6$$

To find some of the critical points, we set $\frac{df}{dx} = 0$ and solve for the roots. We find these roots with Derive by executing the soLve command. The expression is assumed to be set equal to 0 for this operation. The resulting two roots (critical points) are shown in the following display:

$$3: \quad 3x^2 - 8x - 6$$

$$4: \quad x = \frac{4}{3} - \frac{\sqrt{34}}{3}$$

$$5: \quad x = \frac{\sqrt{34}}{3} + \frac{4}{3}$$

The second derivative $\frac{d^2f}{dx^2}$ can be obtained by taking the derivative of $\frac{df}{dx}$. Simply highlight the expression for $\frac{df}{dx}$ using the arrow keys and issue the Calculus Differentiate command. Simplify performs the differentiation and gives the following result:

```
       d        2
6:    ── (3 x  - 8 x - 6)
      dx

7:     6 x - 8
```

Now use the Manage Substitute command to find the value of the second derivative at the critical points. These values are rather complicated and long to retype, so use the highlight (direction keys) and the **F3** key to obtain the values for the replacement of x. After the first value is substituted, execute approX to get the evaluation of $\frac{d^2 f}{dx^2}$ at the first critical value. The result is negative as shown in the following display:

```
7:    6 x - 8

             ⎡ 4     √34 ⎤
8:    6  ⎢ ─  -  ─── ⎥  - 8
             ⎣ 3      3  ⎦

9:    -11.6619
```

Therefore, this critical point produces a local maximum.

Now we go back to the expression for the second derivative and Manage Substitute the second critical point into the function. Issue the approX command to get

```
              ⎡ √34     4  ⎤
10:    6  ⎢ ───  +  ─ ⎥  - 8
              ⎣  3      3  ⎦

11:   11.6619
```

This time the value is positive and we have a local minimum.

To find the inflection points, the roots of $\frac{d^2 f}{dx^2} = 0$ are needed. Highlight the expression for the second derivative and execute soLve. The result is the inflection point

12: $x = \dfrac{4}{3}$

Let's analyze $g(x)$ in a similar manner. However, we will use a different procedure to set up the derivatives. The second method to find derivatives uses the in-line DIF command. See Section 1.7 for the description of this command. To find $\dfrac{dg}{dx}$, **Author**

`DIF((x^2+1)/(x^2-4x+3),x)`.

The order defaults to first-order when no third argument is entered. Simplify to get:

13: $\dfrac{d}{dx} \dfrac{x^2 + 1}{x^2 - 4x + 3}$

14: $- \dfrac{4(x^2 - x - 1)}{(x^2 - 4x + 3)^2}$

We find the roots of $\dfrac{dg}{dx} = 0$ by executing soLve. When the soLve command is given an expression that is not an equation or an inequality, it returns the zeros of the expression. The result is as follows:

14: $- \dfrac{4(x^2 - x - 1)}{(x^2 - 4x + 3)^2}$

15: $x = \dfrac{1}{2} - \dfrac{\sqrt{5}}{2}$

16: $x = \dfrac{\sqrt{5}}{2} + \dfrac{1}{2}$

The second derivative $\frac{d^2g}{dx^2}$ can be determined with a similar procedure.
Author

`DIF((x^2+1)/(x^2-4x+3),x,2)`.

The 2 in the third argument of this command indicates second order.
Simplify to obtain the following:

$$17: \quad \left[\frac{d}{dx}\right]^2 \frac{x^2+1}{x^2-4x+3}$$

$$18: \quad \frac{4(2x^3 - 3x^2 - 6x + 11)}{(x^2 - 4x + 3)^3}$$

Manage Substitute the critical points into the second derivative one at a time and execute approX to get their values. The following display shows these operations for the two critical points:

$$18: \quad \frac{4(2x^3 - 3x^2 - 6x + 11)}{(x^2 - 4x + 3)^3}$$

$$19: \quad \frac{4\left[2\left[\frac{1}{2} - \frac{\sqrt{5}}{2}\right]^3 - 3\left[\frac{1}{2} - \frac{\sqrt{5}}{2}\right]^2 - 6\left[\frac{1}{2} - \frac{\sqrt{5}}{2}\right] + 11\right]}{\left[\left[\frac{1}{2} - \frac{\sqrt{5}}{2}\right]^2 - 4\left[\frac{1}{2} - \frac{\sqrt{5}}{2}\right] + 3\right]^3}$$

20: 0.260990

$$21: \quad \frac{4\left[2\left[\frac{\sqrt{5}}{2} + \frac{1}{2}\right]^3 - 3\left[\frac{\sqrt{5}}{2} + \frac{1}{2}\right]^2 - 6\left[\frac{\sqrt{5}}{2} + \frac{1}{2}\right] + 11\right]}{\left[\left[\frac{\sqrt{5}}{2} + \frac{1}{2}\right]^2 - 4\left[\frac{\sqrt{5}}{2} + \frac{1}{2}\right] + 3\right]^3}$$

22: -12.2609

We see by the results that $x = 1/2 - \sqrt{5}/2$ is a local minimum, and $x = 1/2 + \sqrt{5}/2$ is a local maximum.

74   Chapter 2. Examples

In order to find the points of inflection, highlight the expression for $\frac{d^2g}{dx^2}$ and execute soLve. The result is

$$23: \quad x = -\frac{5^{1/3}}{2} - \frac{5^{2/3}}{2} + \frac{1}{2}$$

$$24: \quad x = \frac{5^{1/3}}{4} + \frac{5^{2/3}}{4} + \frac{1}{2} + \hat{\imath} \left[\frac{16875^{1/6}}{4} - \frac{675^{1/6}}{4} \right]$$

$$25: \quad x = \frac{5^{1/3}}{4} + \frac{5^{2/3}}{4} + \frac{1}{2} + \hat{\imath} \left[\frac{675^{1/6}}{4} - \frac{16875^{1/6}}{4} \right]$$

Therefore, the real value from expression #23, $x = 5^{1/3}/2 - 5^{2/3}/2 + 1/2$, is an inflection point.

The analysis of $h(x)$ proceeds in a similar manner. Author

$$\boxed{\texttt{DIF(x lnx+sinx,x)}}$$

and Simplify to obtain

$$26: \quad \frac{d}{dx}(x \, LN(x) + SIN(x))$$

$$27: \quad LN(x) + COS(x) + 1$$

Now, we should be able to get the critical points by executing soLve. However, Derive can not solve this transcendental expression exactly, so we will help by changing the mode with Options Precision Approximate. In this mode the soLve command performs the bisection algorithm. Before, we execute the command we need to know a reasonable interval to search. Let's graph the function to obtain that interval. Simply execute Plot Plot to obtain the following plot:

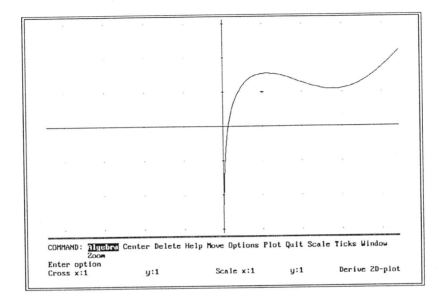

The function crosses the x-axis between 0 and 1. Now reissue soLve. This time the bounds for the bisection algorithm are requested. Use the value of 0 for the lower bound and 1 for the upper bound to obtain

$$27: \quad \text{LN}(x) + \text{COS}(x) + 1$$

$$28: \quad x = 0.136601$$

The second derivative is found by highlighting the first derivative and executing Calculus Differentiate. Then select Simplify to obtain

$$29: \quad \frac{d}{dx}(\text{LN}(x) + \text{COS}(x) + 1)$$

$$30: \quad \frac{1}{x} - \text{SIN}(x)$$

Use the Manage Substitute command to evaluate the second derivative at the critical value. The positive result determines that $x = 0.136601$ is a local minimum.

```
              1
31:      ──────────── - SIN (0.136601)
           0.136601

32:   7.18441
```

Example 2.4 Curvature

The moving power of mathematical invention is not reasoning but imagination.

— Augustus DeMorgan

Subject: Finding the curvature of functions in Cartesian, parametric, and vector form
Reference: Section 1.13
Problem: Find an expression for the curvature and the point with the maximum curvature for the following three functions:

$$y = f(x) = x^5 - x^3 - 4x^2 + 3, \quad -3 \leq x \leq 3$$

$$x = \cos t, \quad y = \sin t, \quad 0 \leq t \leq 2\pi$$

$$\vec{r}(t) = (e^t \cos t)\vec{i} + (e^t \sin t)\vec{j}, \quad 0 \leq t \leq 5$$

Solution: The utility file DIF_APPS.MTH contains commands to find formulas for the curvature of functions. See Section 1.12 for information about the commands in this utility file. For functions in Cartesian form, such as $f(x)$, the CURVATURE command is used. Therefore, the first step in solving this problem is to load the utility file DIF_APPS.MTH. This is done through the execution of the Transfer Load Utility command. Enter DIF_APPS for the file name.

To find an expression for the curvature of $f(x)$, Author the in-line command

CURVATURE(x^5-x^3-4x^2+3,x).

Then Simplify to get the following curvature function:

33: CURVATURE $(x^5 - x^3 - 4x^2 + 3, x)$

34: $\dfrac{2(10x^3 - 3x - 4)}{(25x^8 - 30x^6 - 80x^5 + 9x^4 + 48x^3 + 64x^2 + 1)^{1.5}}$

In this case, we will try to find the point of maximum curvature graphically. The default domain of the plot region should be sufficient since we are only interested in the interval $-3 \leq x \leq 3$. Plot this curvature function by issuing the command Plot Plot. The resulting graph is:

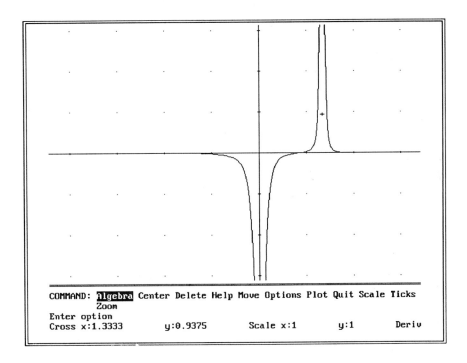

We need to adjust the y-scale in order to plot the parts of the graph we need to see. Reset the scale by choosing the Scale command. Use the **Tab** key to toggle to the y-scale and type 5 and **Enter**. The rescaled plot is drawn as shown.

Chapter 2. Examples

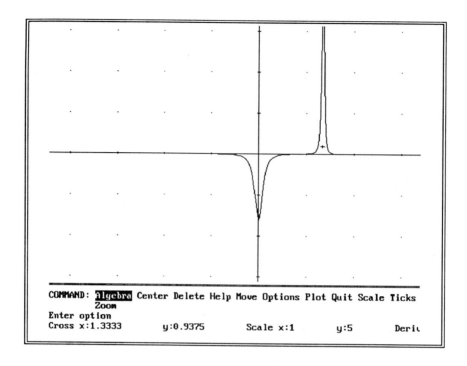

The point where the function reaches its maximum (in this case blows up) can be approximated by using the direction keys to move the cross to its point on the graph. The coordinates of the location of the cross ($x = 1.33333$) are displayed at the bottom of the plot screen shown above.

In order to visualize the relationship between the original function and its computed curvature function, both functions are plotted on the same graph. Return to the Algebra window and highlight the original function. Then issue Plot Plot. The plot of the curvature is redrawn before the new function is plotted. The resulting graph showing both functions reveals the relationship between this function and its curvature.

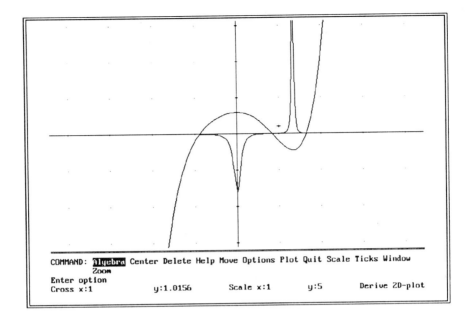

A similar procedure is followed to find the curvature for the second function
$$x = \cos t, \quad y = \sin t, \quad 0 \leq t \leq 2\pi.$$
This time the function is in parametric form so a different command must be used. Author
$$\boxed{\texttt{PARA\_CURVATURE([cost,sint],t)}}.$$
Simplify to get the following curvature function:

```
35:   PARA_CURVATURE ([COS (t), SIN (t)], t)

36:   - SIGN (SIN (t))
```

Set the x-scale to 3 using the Plot Scale command and then plot this function using the Plot command to see the curvature alternates between 1 and -1 over periods of π.

80   Chapter 2. Examples

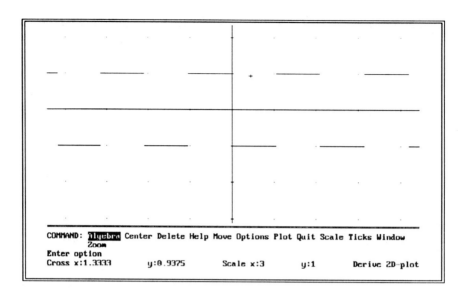

The same procedure is used for the third function, parametric vector equation $\vec{r}(t) = (e^t \cos t)\vec{i} + (e^t \sin t)\vec{j}$. First, Author

$$\boxed{\texttt{PARA\_CURVATURE([Alt-e\^{}t cost,Alt-e\^{}t sint],t)}}$$

Execute Simplify to obtain

```
37:  PARA_CURVATURE ([e^t COS (t), e^t SIN (t)], t)

38:  0.707106 e^{-t} SIGN (COS (t) - SIN (t))
```

Set the x-scale $= 1$ and y-scale $= 0.5$ Using Plot Scale. Move the cross to $x = 2.5$ and $y = 0$ and execute Center. Plot the curvature function with the Plot command to produce the graph as follows:

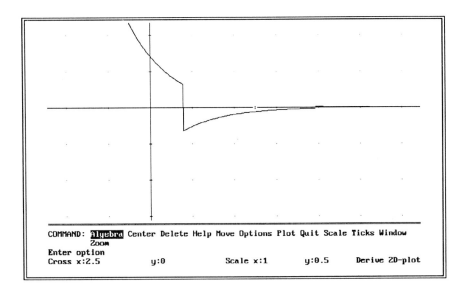

Example 2.5 Root Finding

A great discovery solves a great problem but there is a grain of discovery in the solution of any problem.

— George Polya [1946]

Subject: Finding or approximating roots of functions
References: Sections 1.9 and 1.10
Problem: Find all the real roots of the following functions:

$$f(x) = x^3 + 4x^2 - .6x + 3.7$$
$$g(x) = x^5 + 7x^4 - 0.2x^3 - x^2 + 0.5x - 10$$
$$h(x) = e^x - x \sin x - 2$$

Solution: Finding the roots of some equations can be quite simple. First, enter the equation into the work area with the Author command. For $f(x)$, the expression is entered as x^3+4x^2-0.6x+3.7 . Then just execute soLve. The result is:

Chapter 2. Examples

$$1: \quad x^3 + 4x^2 - 0.6x + 3.7$$

$$2: \quad x = -\left[\frac{\sqrt{11025345}}{900} + \frac{499}{108}\right]^{1/3} - \left[\frac{499}{108} - \frac{\sqrt{11025345}}{900}\right]^{1/3} - \frac{4}{3}$$

$$3: \quad x = \left[\frac{\sqrt{11025345}}{7200} + \frac{499}{864}\right]^{1/3} + \left[\frac{499}{864} - \frac{\sqrt{11025345}}{7200}\right]^{1/3} - \frac{4}{3} + \hat{\imath}\left[\frac{254853}{172800}\right]$$

$$4: \quad x = \left[\frac{\sqrt{11025345}}{7200} + \frac{499}{864}\right]^{1/3} + \left[\frac{499}{864} - \frac{\sqrt{11025345}}{7200}\right]^{1/3} - \frac{4}{3} + \hat{\imath}\left[\frac{499\sqrt{1}}{115}\right]$$

As expected, this cubic polynomial has three roots. However, in this case only one is real. The other two are complex, and all three are messy. The two complex roots are truncated in the figure and on the screen. The direction arrow keys can be used to see the portion of the expression off the right of the screen. Highlight each root in turn and execute `approX`. The three approximate roots in decimal form are as follows:

$$5: \quad x = -4.33526$$
$$6: \quad x = 0.167632 - 0.908494\,\hat{\imath}$$
$$7: \quad x = 0.167632 + 0.908494\,\hat{\imath}$$

Now we will try the same procedure for the fifth-degree polynomial $g(x)$. Author

`x^5+7x^4-0.2x^3-x^2+0.5x-10`

and execute `soLve`. This time Derive doesn't help us.

Before turning to an approximate numerical procedure, we may be able to approximate some of the roots using graphics. Move the highlight to the expression for $g(x)$ through the use of the direction arrow keys (↑ and ↓). Execute `Plot Scale` and set x-scale $= 2$ and y-scale $= 5$. Then execute `Plot`. The resulting plot with the designated scale parameters is:

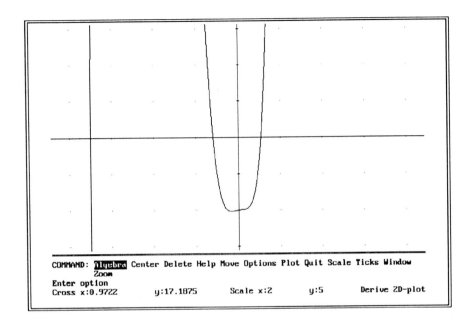

Carefully move the cross to a point near the three x-intercepts or roots. We come up with the following three approximate values for the roots: -7, -1.2, and 1.1. Now change the mode from Exact to Approximate using the Options Precision command. Set the accuracy to 6 digits. Now the soLve command performs the bisection approximation procedure. Return the algebra window and execute soLve. To find the root near $x = -7$ provide lower and upper values of -8 and -6, respectively. The result is

```
8:    x^5 + 7 x^4 - 0.2 x^3 - x^2 + 0.5 x - 10

9:    x = -7.00256
```

Repeat this procedure to find the other two roots using the bisection search intervals of $(-1.5, -1)$ and $(0.9, 1.1)$. The results are as follows:

```
9:    x = -7.00256
10:   x = -1.19065
11:   x = 1.07710
```

Finding all the real roots of $h(x)$ is also difficult. Author Alt-e^x-x-2. Issue Plot Plot to obtain

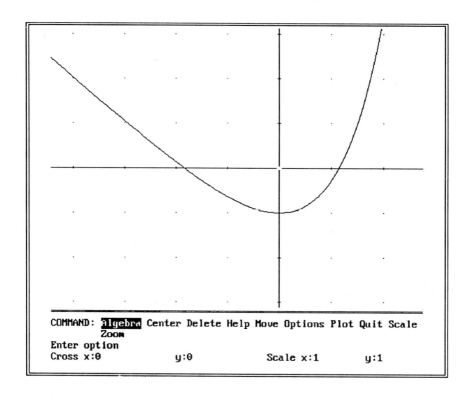

We see that $h(x)$ has one positive and one negative root. Return to the Algebra window. Change the precision to Approximate using the Options Precision command. Then execute soLve. In order to obtain the negative root, input −4 for the lower bound and 0 for the upper bound. To obtain the positive root execute soLve with lower bound of 0 and upper bound of 4. The results of these operations are as follows:

12: $e^x - x - 2$

13: $x = -1.84140$

14: $x = 1.14619$

Example 2.6 Integration

The union of the mathematician with the poet, fervor with measure, passion with correctness, this surely is the ideal.

— William James [1879]

Subject: Evaluation of definite and indefinite integrals
References: Sections 1.7 and 1.13
Problem: Compute the following integrals:

$$\int x^2 e^x dx$$

$$\int_0^2 \ln(x^2 + 1)dx$$

$$\int_0^{2\pi} \cos(x^2)dx$$

Approximate the integral (area) in the interval $0 < x < 1$ of the function represented by the following table of data points:

x	y
0.0	3
0.2	4
0.4	5
0.5	2
0.6	2.1
0.9	3
1.0	2.9

Solution: Integrals are set up with the Calculus Integrate menu command or with the in-line INT command. Using the first method for the first integral, Author $\boxed{\texttt{x\textasciicircum2 Alt-e\textasciicircum x}}$. Then select, Calculus Integrate. Answer the menu queries as follows: variable of integration: x; lower limit: *(leave blank)*; upper limit: *(leave blank)*.

The blank responses for the limits indicate this is indefinite integration or antidifferentiation. Execute Simplify. The set-up integral and the result of the integration are as follows:

$$2:\quad \int x^2 \hat{e}^x\, dx$$

$$3:\quad \hat{e}^x (x^2 - 2x + 2)$$

Notice the answer does not contain an arbitrary constant of integration. Derive does not automatically add in the constant. Let's check ourselves and Derive by taking the derivative of this expression to see if it is equivalent to the integrand. Just select Calculus Differentiate and respond with differentiation with respect to x and an order of 1. Simplify to obtain the result we desired:

$$2:\quad \int x^2 \hat{e}^x\, dx$$

$$3:\quad \hat{e}^x (x^2 - 2x + 2)$$

$$4:\quad \frac{d}{dx} \hat{e}^x (x^2 - 2x + 2)$$

$$5:\quad x^2 \hat{e}^x$$

We will find the second integral

$$\int_0^2 \ln(x^2 + 1)\,dx$$

with the second method for integration of using the in-line INT command. Author

$$\boxed{\text{INT}(\ln(x^2+1),x,0,2)}.$$

This is a definite integral so the limits of integration are placed in the third and fourth arguments. If this had been an indefinite integral, these parameters would have been omitted. Simplify to obtain the following result:

$$6: \quad \int_0^2 \ln(x^2 + 1)\, dx$$

$$7: \quad -2\,\text{ATAN}\left[\frac{1}{2}\right] + 2\,\text{LN}\,(5) + \pi - 4$$

This is the exact solution found by symbolic integration. To obtain an approximation to these real numbers, select the approX menu command which produces the following decimal approximation:

$$7: \quad -2\,\text{ATAN}\left[\frac{1}{2}\right] + 2\,\text{LN}\,(5) + \pi - 4$$

$$8: \quad 1.43317$$

Of course, the third integral

$$\int_0^{2\pi} \cos(x^2)\,dx$$

can be set up using either method. The in-line set-up command is *Authored* by

$$\boxed{\texttt{INT(cos(x\^{}2),x,0,2pi)}}.$$

This time Simplify doesn't help. Derive just returns the original set-up for the integral.

$$9: \quad \int_0^{2\pi} \cos(x^2)\, dx$$

$$10: \quad \int_0^{2\pi} \cos(x^2)\, dx$$

The reason for this is that Derive cannot integrate this integrand exactly. However, Derive can still help. Execute approX to obtain an adaptive quadrature approximation to the definite integral. The approximate value is as shown in the following display:

$$10: \int_0^{2\pi} \cos(x^2)\, dx$$

$$11:\ 0.704683$$

The accuracy goal in the form of significant digits for this numerical approximation is controlled by the Options Precision menu command. Execute the command and enter the number of digits of accuracy desired. Thereafter, Derive tries to achieve that level of accuracy by approximating the error in their numerical technique. If Derive is doubtful that it obtained the requested accuracy, it gives a warning of 'Dubious Accuracy'.

In order to approximate integrals of functions represented by discrete data using the trapezoidal rule, the INT_DATA command from the utility file NUMERIC is used. Execute Transfer Load Utility for NUMERIC. To use the command the data must be in a matrix with 2 columns.

For our data file

x	y
0.0	3
0.2	4
0.4	5
0.5	2
0.6	2.1
0.9	3
1.0	2.9

execute Declare Matrix with 7 rows and 2 columns. Type in the data points for the appropriate entries. The result is:

$$12: \begin{bmatrix} 0 & 3 \\ 0.2 & 4 \\ 0.4 & 5 \\ 0.5 & 2 \\ 0.6 & 2.1 \\ 0.9 & 3 \\ 1 & 2.9 \end{bmatrix}$$

Then Author $\boxed{\text{INT\_DATA \#12}}$. Simplify to obtain the matrix

$$14: \begin{bmatrix} 0 & 0 \\ \frac{1}{5} & \frac{7}{10} \\ \frac{2}{5} & \frac{8}{5} \\ \frac{1}{2} & \frac{39}{20} \\ \frac{3}{5} & \frac{431}{200} \\ \frac{9}{10} & \frac{73}{25} \\ 1 & \frac{643}{200} \end{bmatrix}$$

The matrix shows the approximate antiderivative function for the x points in the data set. For the integral over the entire interval $0 < x < 1$, the last result is used. Therefore, the approximation to the integral or area for our problem is $643/200$.

Example 2.7 Arc Length

Mathematics ... gives life to her own discoveries; she awakens the mind and purifies the intellect; she brings light to our intrinsic ideas; she abolishes the oblivion and ignorance which are ours by birth.

— Proclus [Fifth Century]

Subject: Finding the arc length of functions
Reference: Section 1.13
Problem: Find the arc length of the curve of the graph for the following functions in the specified intervals:

$$y = \cos x, \quad 0 \leq x \leq \pi$$

$$x = \cos t \quad \text{and} \quad y = \sin t \quad \text{with} \quad 0 \leq t \leq 2\pi.$$

Solution: The utility file INT_APPS.MTH contains commands to find the arc length of functions. See Section 1.13 for information about this and other commands in this utility file. For functions in Cartesian form the ARC_LENGTH command is used. Therefore, the first step in solving this problem is to load the utility file INT_APPS.MTH. This is done through the execution of the `Transfer Load Utility` command. Enter `INT_APPS` for the file name.

To find the arc length of the first function $\cos x$, Author the in-line command

`ARC_LENGTH(cosx,x,0,pi)`.

Then `Simplify` to get the result:

```
22:    ARC_LENGTH (COS (x), x, 0, π)

            π
23:    ∫    √(SIN (x)² + 1) dx
            0
```

This is not what we desired. It is the integral set-up for the arc length but not the numerical measure. Derive is telling us it cannot perform this integration exactly. So to get the answer, we execute `approX` to numerically evaluate the definite integral. Now we get the type of answer we expected

$$23: \int_0^\pi \sqrt{(\text{SIN}(x)^2 + 1)}\, dx$$

24: 3.82020

Does this answer make sense? The $\cos x$ function on our interval goes from $(0,1)$ to $(\pi, -1)$. The straight line distance between those two points is $\sqrt{\pi^2 + 2^2}$. Let's use Derive to evaluate this. Author the expression

sqrt(pi^2+2^2)

and approX. The result is

$$25: \sqrt{(\pi^2 + 2^2)}$$

26: 3.72419

The curved graph of the cosine function is just a little longer than the straight-line path between the two end points.

To find the arc length for the second function, which is a circle of radius 1 given in parametric form, a similar procedure is used. The function of interest is

$$x = \cos t \quad \text{and} \quad y = \sin t \quad \text{with} \quad 0 \le t \le 2\pi.$$

This time the command to Author for the set-up of the arc length is

PARA_ARC_LENGTH([cost,sint],t,0,2pi).

Then execute the Simplify command. This time Derive can do the integration symbolically with the result as expected

27: PARA_ARC_LENGTH ([COS (t), SIN (t)], t, 0, 2 π)

28: 2 π

This is just the formula for the circumference of the circle with unit radius.

Can Derive help us if we had a general circle of radius r? This changes the parametric equation to

$$x = r\cos t \quad \text{and} \quad y = r\sin t \quad \text{with} \quad 0 \le t \le 2\pi.$$

To find the arc length of the circle of radius r, Author the in-line command
$$\boxed{\texttt{PARA\_ARC\_LENGTH([rcost,rsint],t,0,2pi)}}.$$
Select Simplify to obtain

```
29:   PARA_ARC_LENGTH ([r COS (t), r SIN (t)], t, 0, 2 π)
30:   2 π |r|
```

As we can see, Derive handles the symbolic parameter r without any difficulty.

Example 2.8 Projectile Motion

Come one, come all! this rock shall fly From its firm base as soon as I.

— Sir Walter Scott [1810]

Subject: Projectile motion
References: Sections 1.7 and 1.9
Problem: A projectile is fired over horizontal ground at an initial speed of v meters/second at an angle of elevation a. a is between 0 and $\pi/2$ radians. Where will the projectile be t seconds later? Assume the only force acting on the projectile is gravity that produces a downward acceleration of 9.8 meters/second^2.

Solution: First we let $x(t)$ represent the horizontal displacement and $y(t)$ represent the vertical. We make the firing point the origin (0,0). To start our calculations let's use a special case of $a = \pi/2$ radians (90°). This is the case of the projectile firing straight up, and the problem reduces to one space dimension y.

Start with acceleration $y'' = -9.8$. Integrate this to find the velocity $y'(t) = -9.8t + c_1$ (You shouldn't need to use Derive for this easy integration). Since $y'(0) = v$, $c_1 = v$ and $y'(t) = -9.8t + v$. The vertical displacement y is found by another integration. This time use Derive and Author $\boxed{\texttt{INT(-9.8t+v,t)}}$. Simplify to obtain the following result:

$$1: \quad \int (-9.8\,t + v)\,dt$$

$$2: \quad t\,v - \frac{49\,t^2}{10}$$

Remember, Derive does not provide a constant of integration, so $y = -4.9t^2 + vt + c_2$. Since the firing takes place at the origin, $y(0) = 0$ which makes $c_2 = 0$. Therefore, the height of the projectile at time t is simply $y(t) = -4.9t^2 + vt$.

Now, we will use vector equations to solve the problem for any angle a between 0 and $\pi/2$ radians. Start with the acceleration due to gravity in vector form $[x'', y''] = [0, -9.8]$. Integrate to find the velocity vector $[c_1, -9.8t + c_2]$. Once again, this is done without Derive. The initial velocity vector is $[v \cos a, v \sin a]$ which makes $c_1 = v \cos a$ and $c_2 = v \sin a$. Therefore, the velocity vector $[x', y']$ is $[v \cos a, -9.8t + v \sin a]$. To find the displacement vector $[x, y]$, we need to integrate the velocity. Author

> INT([v cosa, -9.8t+v sin a],t)

and Simplify. The result is

$$3: \quad \int [v\,\cos(a),\, -9.8\,t + v\,\sin(a)]\,dt$$

$$4: \quad \left[t\,v\,\cos(a),\, t\,v\,\sin(a) - \frac{49\,t^2}{10}\right]$$

Add the constants of integration, b and c, to the respective components of the vector. This can be done through the Author command by either retyping the expression with the appropriate constants or by using the **F3** key to bring the highlighted expression into the author line of the menu and editing it. In order to edit an expression in the author line, we use the **Ctrl-s** keys to move the cursor to the left and the **Ctrl-d** keys to move the cursor to the right. The **Ins** key can be used to toggle on and off the insert mode. Press the **Enter** key when the expression is completed. The work area now contains the following expressions:

Chapter 2. Examples

$$5: \quad \left[t\, v\, \cos(a) + b,\ t\, v\, \sin(a) - \frac{49\, t^2}{10} + c \right]$$

First we use `Manage Substitute` to replace t with 0 and `Simplify`. Then, since the initial displacements were (0,0), we issue the `soLve` command. The result is the following trivial values for these two constants:

$$6: \quad \left[0\, v\, \cos(a) + b,\ 0\, v\, \sin(a) - \frac{49\, \cdot 0^2}{10} + c \right]$$

7: [b, c]

8: [b = 0, c = 0]

Therefore, the displacement vector in meters of a projectile fired at an angle a is $[(v \cos a)t, -4.9t^2 + (v \sin a)t]$. Let's plot this parametric vector for several values of a for $0 < t < 10$. We choose $v = 100$ and $a = \pi/8, \pi/4, 3\pi/8$, and $9\pi/20$. We have to plot these one at a time, going back and forth from the algebra window to the plot window.

First, we `Author`

`[(100cos a)t, -4.9t^2+(100sin a)t]`.

Then select `Manage Substitute` with replacement of a with `pi/8` and just press `Enter` when queries for a value of t. Then select the `Plot` menu. Before plotting the function we need to rescale. Select `Scale` and set x-scale = 200 and y-scale = 200. Issue `Plot`. Derive realizes the highlighted function is in parametric form, so it produces a submenu to set up several parametric plotting values. Since we want $0 < t < 10$, set `Min` at 0 and `Max` at 10 and press `Enter`. The resulting plot is as follows:

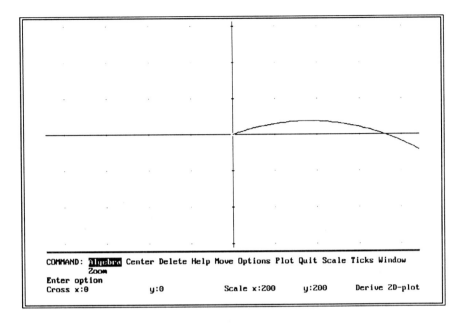

In order to produce each of the other three plots on the same graph, select Algebra and repeat the entire process using the remaining 3 different values of *a*. Each time the Plot menu is selected, all the previous plots will be produced before the new plot is drawn. This is a time consuming process and one of the limitations of Derive. The resulting plot screen is as follows:

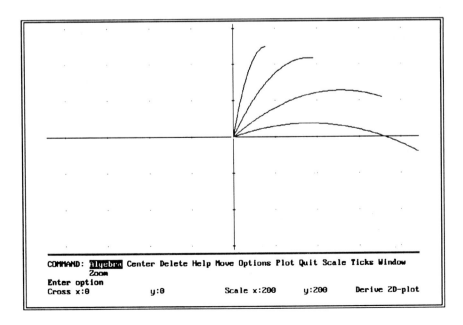

The resulting algebra work area is as follows:

```
11:  [100 COS [π/8] t, -4.9 t² + 100 SIN [π/8] t]

12:  [100 COS [π/4] t, -4.9 t² + 100 SIN [π/4] t]

13:  [100 COS [3π/8] t, -4.9 t² + 100 SIN [3π/8] t]

14:  [100 COS [9π/20] t, -4.9 t² + 100 SIN [9π/20] t]
```

Example 2.9 Radioactive Decay

These methods of learning about nature are increasingly more important in more and more fields. They also underlie the process by which engineers create the technologies that exercise vast influence over all our lives..

— Samuel Goldberg, *UME Trends* [1990]

Subject: Using radioactive decay of Carbon-14 to determine the age of paintings

Reference: Section 1.14

Problem: Carbon-14 has a half-life of about 5700 years. In other words, 50% of Carbon-14 decays into Nitrogen during its half-life. The rate of decay of a radioactive element like Carbon-14 is proportional to the amount of the element present. If $a(t)$ is the amount of Carbon-14, the model that describes its decay is

$$\frac{da}{dt} = -ka, \text{ with } a(0) = c.$$

If a painting has between 90% and 91% of its original Carbon-14, how many years ago was the painting created?

Solution: The proportionality constant k is not known. However, from the half-life information, we know $a(5700) = 0.5a(0) = 0.5c$. By rewriting the equation in the form

$$\frac{da}{dt} + ka = 0,$$

the equation is seen to be linear and the appropriate arguments for the LINEAR1 command can be determined (See Section 1.12).

To use this command Transfer Load Utility the utility file ODE1. Then Author the expression

$$\boxed{\text{LINEAR1(k,0,t,a,5700,0.5c)}}.$$

Execute Simplify to obtain

$$
\begin{array}{l}
27: \quad \text{LINEAR1 (k, 0, t, a, 5700, 0.5 c)} \\[1em]
28: \quad a = \dfrac{c\,e^{k(5700 - t)}}{2}
\end{array}
$$

Now substitute the initial condition $a(0) = c$ into the solution. Use the Manage Substitute command to do this. We want to substitute 0 for t and c for a and leave the unknowns c and k as they are. Just press **Enter** when Derive asks for what to substitute for c and k. The result is

$$
\begin{array}{l}
28: \quad a = \dfrac{c\,e^{k(5700 - t)}}{2} \\[1em]
29: \quad c = \dfrac{c\,e^{k(5700 - 0)}}{2}
\end{array}
$$

We want to find k, so ask Derive to soLve for the solve variable k. Derive tries to do this but can not solve this exponential equation. We can help by taking the logarithm of the entire equation before asking Derive to soLve. Highlight the equation (expression #29) and Author $\boxed{\text{ln(F4)}}$. The execute soLve and approX. The result is what we wanted.

Chapter 2. Examples

$$30: \quad \text{LN}\left[c = \frac{c\, \hat{e}^{k\, 5700}}{2}\right]$$

$$31: \quad k = \frac{\text{LN}(2)}{5700}$$

$$32: \quad k = 1.21605\ 10^{-4}$$

Now that we know the value of k for Carbon-14, we can find the values of t where $a = 0.90c$ and $a = 0.91c$. Highlight the solution function (expression #28) and use the Manage Substitute command to replace k with its value $\ln(2)/5700$. Then Simplify to obtain the equation

$$34: \quad a = \frac{c\, \hat{e}^{\text{LN}(2)\,/\,5700\,(5700\,-\,t)}}{2}$$

$$35: \quad a = 2^{-t/5700}\, c$$

Select Manage Substitute for this expression to replace a with 0.9 and c with 1. Try executing soLve with a solve variable of t. Since this equation is similar to the previous exponential equation, Derive can not solve this one either. This time, let's approximate a solution using the bisection method. To start this method we need to know the end points of an interval that includes the solution. In this case -1.21605×10^{-4} is small, so t must be quite large (near 10^4) to satisfy the equation. Let's try the interval $500 < t < 2000$. If our interval is not appropriate, Derive will indicate this by printing No solution found. Another method of estimating the solution would be to graph the function. In this case we won't produce the graph.

To implement the bisection algorithm, we issue Options Precision Approximate. The required digits of accuracy is set to 6. Execute soLve with the parameters Lower: 500 and Upper: 2000. The result of the approximation is

```
                   - t/5700
36:   0.9 = 2              1
37:   t = 866.416
```

The same method is used for $a = 0.91c$. The resulting display is as follows:

```
                   - t/5700
38:   0.91 = 2             1
39:   t = 775.548
```

We have determined the painting is between 775 and 866 years old.

Example 2.10 Area and Volume

All men by nature desire Knowledge.

— Aristotle [4th century B.C.]

Subject: Finding the area and volume of shapes defined by functions
References: Sections 1.9 and 1.13
Problem: Given the functions

$$y = f_1(x) = \cos x$$
$$y = f_2(x) = \sin x$$
$$y = f_3(x) = x^2$$
$$z = g(x, y) = e^{x+y} ,$$

find the area of the regions between f_1 and f_2, f_1 and f_3, and f_2 and f_3, between their intersections, and then find the volume of the solid formed by the function $g(x, y)$ above the region formed by $f_1(x)$ and $f_3(x)$.

Solution: We will use integration to find the areas. In order to get the limits of integration, we need to find the points where these functions intersect. Let's get some geometric intuition about these regions by plotting $f_1(x), f_2(x),$ and $f_3(x)$ on the same axes.

An efficient way to graph 3 functions at once is to enter the functions as components of a vector and graph the vector. Therefore, we will enter each of the 3 functions as expressions into the work area. To do this, Author sinx, press **Enter**, Author cosx, press **Enter**, and Author x^2, press **Enter**. The resulting display is as follows:

```
1:    SIN (x)

2:    COS (x)

          2
3:    x
```

Next, form a vector by placing the appropriate expression numbers as components in a vector (For example, [#m,#n,#p], where m, n and p are the Derive statement numbers for the functions). For the Derive work area in our example, Author [#1,#2,#3]. To plot these functions, we execute Plot Plot. The result using the scale graphing parameters of x-scale $= 1.3$ and y-scale $= 0.5$ is as shown.

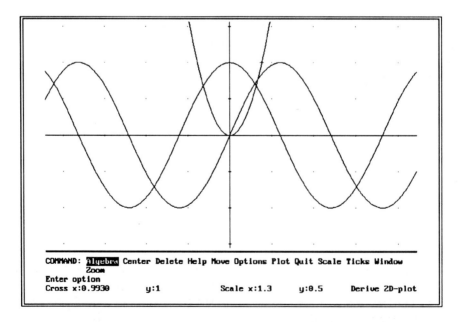

Let's label each of the six different types of regions and proceed to find the area of each of them. The following graph has the regions numbered

1-6 by hand. Unfortunately, Derive does not have a capability to label graphs. This is one of the limitations of Derive discussed in Section 1.14.

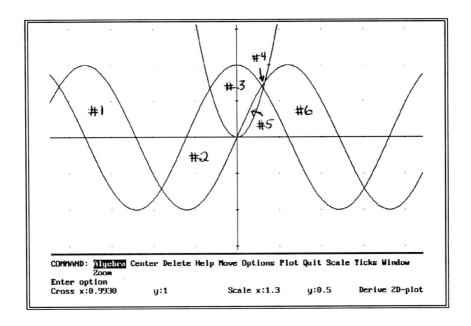

We will start with region #1. By moving the movable cross with the direction arrow keys, we see that two intersection points of the top curve $\sin x$ and the bottom curve $\cos x$ are $x \approx -6$ and $x \approx -2$. When we Author $\boxed{\text{sinx-cosx}}$ and execute soLve in the Exact mode, we get $x = \pi/4$ as shown in the following display:

```
5:   SIN (x) - COS (x)

             π
6:   x = ───
             4
```

This principal root ($\pi/4$) is actually 2π and π larger than the two roots in the interval we are searching. This is verified by changing the mode through the Options Precision Approximate command. Performing bisection by executing soLve with intervals of $(-7, -5)$ and $(-3, -2)$ results in -5.49787 and -2.35623, the decimal equivalents of $-7\pi/4$ and $-3\pi/4$.

Therefore, the area of region #1 is

$$\int_{-\frac{7\pi}{4}}^{-\frac{3\pi}{4}} (\sin x - \cos x)dx.$$

Before integrating, return to the Exact mode by executing Options Precision Exact. Then highlight the expression $\sin x - \cos x$, and issue Calculus Integrate. The variable of integration is x, for the lower limit enter $\boxed{-7\text{pi}/4}$, and for the upper limit enter $\boxed{-3\text{pi}/4}$. Perform the integration by executing Simplify. The area of the region is as follows:

```
              - 3 π / 4
9:        ∫              (SIN (x) - COS (x)) dx
              - 7 π / 4

10:   2 √2
```

There are an infinite number of regions with the same area as #1.

Finding the area of region #2 is slightly more complicated. In the approximate interval $\frac{-3\pi}{4} < x < -0.8$, the integrand is $\cos x - \sin x$; then from $-0.8 < x < 0$, the integrand is $x^2 - \sin x$. We need to find the exact intersection point of x^2 and $\cos x$. Author $\boxed{\text{cosx-x}^2}$. Issue Options Precision Approximate and soLve. Use a lower bound of -10 and an upper bound of 0. The command and its result are

```
                         2
11:    COS (x) - x

12:    x = -0.824132
```

Therefore, the area of region #2 is

$$\int_{-\frac{3\pi}{4}}^{-0.824132} (\cos x - \sin x)dx + \int_{-0.82413}^{0} (x^2 - \sin x)dx.$$

Execute the Author command and enter

$\boxed{\text{INT(cosx -sinx,x,-3pi/4,-0.824132) +INT(x}^\wedge\text{2-sinx,x,-0.824132,0)}}$.

Issue `Simplify` to obtain the following the result for the area of region #2:

$$13: \int_{-3\pi/4}^{-0.824132} (\cos(x) - \sin(x)) \, dx + \int_{-0.824132}^{0} (x^2 - \sin(x)) \, dx$$

$$14: \quad 1.86683$$

Finding the area of region #3 is similar to the procedure for the area of region #2. In the interval $-0.824132 < x < 0$, the integrand is $\cos x - x^2$; then in the interval $0 < x < \frac{\pi}{4}$, the integrand is $\cos x - \sin x$. Therefore, the area of region #3 is

$$\int_{-0.824132}^{0} (\cos x - x^2) dx + \int_{0}^{\frac{\pi}{4}} (\cos x - \sin x) dx.$$

In order to determine this area, evaluate the integral by *Authoring* $\boxed{\text{INT(cosx-x\^{}2,x,-0.824132,0) +INT(cosx-sinx,x,0,pi/4)}}$.
Issue `Simplify` to obtain the command and its result as follows:

$$15: \int_{-0.824132}^{0} (\cos(x) - x^2) \, dx + \int_{0}^{\pi/4} (\cos(x) - \sin(x)) \, dx$$

$$16: \quad 0.96159$$

Finding the area of the very small region #4 is similar to the previous procedures. In the interval $\frac{\pi}{4} < x < 0.824132$, the integrand is $\sin x - \cos x$. The intersection of $\sin x$ and x^2 must be determined to find the next interval of integration. Author $\boxed{\text{sinx-x\^{}2}}$. Execute `Simplify` to obtain

$$17: \quad \sin(x) - x^2$$

$$18: \quad x = 0.876726$$

Therefore, the interval of integration is $0.824132 < x < 0.876726$ and the integrand is $\sin x - x^2$. The formula for the area of region #4 is

$$\int_{\frac{\pi}{4}}^{0.824132} (\sin x - \cos x)dx + \int_{0.824132}^{0.876726} (\sin x - x^2)dx.$$

In order to determine this area, evaluate this integral by *Authoring*

`INT(sinx-cosx,x,pi/4,0.824132) +INT(sinx-x^2,x,0.824132,0.876726)`.

Simplify to obtain the area of:

```
                0.824132                          0.876726
               ⌠                                  ⌠              2
      19:      ⌡    (SIN (x) - COS (x)) dx  +     ⌡    (SIN (x) - x ) dx
                π/4                                0.824132

      20:  0.00253425
```

This is getting to be routine. The area of region #5 is

$$\int_0^{\frac{\pi}{4}} (\sin x - x^2)dx + \int_{\frac{\pi}{4}}^{0.824132} (\cos x - x^2)dx.$$

In order to determine this area, evaluate the integral by *Authoring*

`INT(sinx-x^2,x,0,pi/4) +INT(cosx-x^2,x,pi/4,0.824132)`.

Simplify to get the area of region #5 as follows:

```
                π/4                              0.824132
               ⌠              2                  ⌠              2
      21:      ⌡    (SIN (x) - x ) dx  +         ⌡    (COS (x) - x ) dx
                0                                 π/4

      22:  0.133163
```

Finally, the area of region #6 is

$$\int_{0.824132}^{0.876726} (x^2 - \cos x)dx + \int_{0.876726}^{\frac{5\pi}{4}} (\sin x - \cos x)dx.$$

In order to determine this area, evaluate the integral by *Authoring*

`INT(x^2-cosx,x,0.824132,0.876726) +INT(sinx-cosx,x,0.876726,5pi/4)`.

Simplify to get the area as follows:

$$23: \int_{0.824132}^{0.876726} (x^2 - \cos(x))\, dx + \int_{0.876726}^{5\pi/4} (\sin(x) - \cos(x))\, dx$$

$$24:\ 2.82589$$

Now, let's tackle the volume problem. We need to find the volume of the solid formed by the function e^{x+y} above the region between $\cos x$ and x^2. From what we already determined the intersection points of $\cos x$ and x^2 are -0.824132 and 0.824132. So one way to find this volume is to calculate

$$\int_{-0.824132}^{0.824132} \int_{x^2}^{\cos(x)} e^{x+y}\, dy\, dx.$$

To perform this integration with Derive, Author

$$\boxed{\texttt{INT(INT(Alt-e\textasciicircum(x+y),y,x\textasciicircum2,cosx), x,-0.824132,0.824132)}}\ .$$

Then Simplify to obtain

$$25: \int_{-0.824132}^{0.824132} \int_{x^2}^{\cos(x)} e^{x+y}\, dy\, dx$$

$$26: \int e^{\cos(x)+x}\, dx + \hat{\imath}\, e^{-206033/250000}\, dx + \hat{\imath}\, e^{-1/4} \left[\sqrt{\pi}\, \mathrm{ERF}\left[-\dfrac{331033\,\hat{\imath}}{250000}\right] \right] \Big/ 2 + \sqrt{\pi}\, E$$

This mess of special functions and imaginary numbers motivate us to try a numerical quadrature to find the real value of the integral. Highlight the integral expression and execute **approX**. Now we get the real numerical result of 2.05987 that we desired.

There is a second way to obtain this volume. The utility file INT_APPS has a command VOLUME designed for this purpose. Load the utility file with the `Transfer Load Utility` command. Author the expression

$$\boxed{\texttt{VOLUME(x,-0.824132,0.824132,y,x\textasciicircum2,cosx,z,0,Alt-e\textasciicircum(x+y))}}\ .$$

This time use the approX command right away to obtain the same answer as shown:

```
28:  VOLUME (x, -0.824132, 0.824132, y, x², COS (x), z, 0, e^(x+y))
29:  2.05987
```

Example 2.11 Polar Equations

"What's the good of Mercator's North Poles and Equators, Tropics, Zones, and Meridian Lines?"
So the Bellman would cry: and the crew would reply,
"They are merely conventional signs!"

— Lewis Carroll, *The Hunting of the Snark* [1876]

Subject: Plotting and using polar equations
References: Sections 1.9 and 1.13
Problem: Given the polar equations

$$r^2 = 4\cos 2\theta \quad \text{and}$$

$$r = 1 - \cos\theta, \quad \text{with} \quad -\pi < \theta < \pi,$$

plot the curves, find their points of intersection, their Cartesian derivatives $\frac{dy}{dx}$, and the formulas for their curvature.

Solution: In order to plot the equation $r^2 = 4\cos 2\theta$, it must be entered as two polar functions, $r = \pm 2\sqrt{\cos 2\theta}$. Because of this there are a total of three functions to investigate. For ease of plotting, we assemble the three functions as components of a vector. Enter the vector into the work area through the Author of

[2 sqrt(cos(2t)), -2 sqrt(cos(2t)), 1-cost]

The variable t is used instead of θ for convenience. θ could be entered using the **Alt-h** keys. To plot these functions, select the Plot submenu. Change the type of plot from its default of Cartesian to polar using the Options Type Polar selection. This command changes the display to the following polar plotting screen:

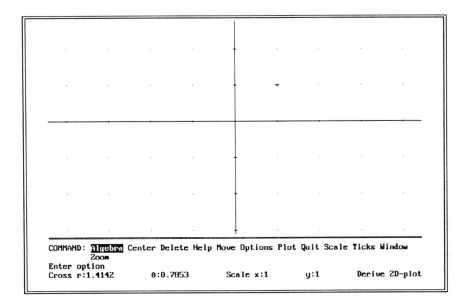

Select Plot and set Min to $-\pi$ and Max to π. The resulting plot on the default x and y scaled axes is:

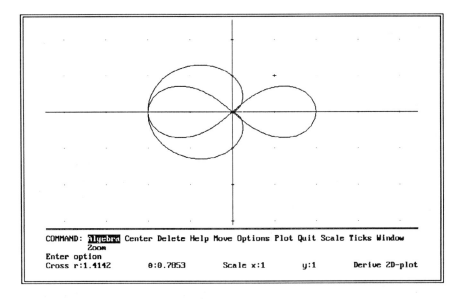

We see that the second branch of the equation $r^2 = 4\cos 2\theta$ is a repeat of the first in this interval of θ. Therefore, from now on, we will only work with the two functions $r = 2\sqrt{\cos 2\theta}$ and $r = 1 - \cos\theta$.

Two obvious points of intersection are the origin and $x = -2, y = 0$ ($r = 2, \theta = -\pi$). To find the other two intersection points, return to the Algebra window and Author the expression

$$\boxed{\text{2 sqrt(cos(2t)) - 1 + cost}}.$$

Set Options Precision to Approximate and execute soLve. Input the first interval for bisection as $-3 < t < 0$. The result is:

```
2:    2 √(COS (2 t)) - 1 + COS (t)
3:    t = -0.775193
```

Highlight the expression (#2) and repeat soLve with $0 < t < 3$ to get the symmetric positive value.

```
2:    2 √(COS (2 t)) - 1 + COS (t)
3:    t = -0.775193
4:    t = 0.775193
```

To find the r values for these two values of t (actually θ), Author $\boxed{\text{1-cost}}$ and Manage Substitute the two values in for t. Simplify each expression to get the following results:

```
5:    1 - COS (t)
6:    1 - COS (0.775193)
7:    0.285714
8:    1 - COS (-0.775193)
9:    0.285714
```

Using $x = r\cos\theta$ and $y = r\sin\theta$, the following operations determine the x and y values for the two intersection points.

```
10:   x = 0.285714 COS (-0.775193)
11:   x = 0.204081
12:   y = 0.285714 SIN (-0.775193)
13:   y = -0.199958
14:   x = 0.285714 COS (0.775193)
15:   x = 0.204081
16:   y = 0.285714 SIN (0.775193)
17:   y = 0.199958
```

Therefore, the four intersection points are: i) the origin; ii) $x = 2$, $y = 0$, $r = 2$, $\theta = -\pi$; iii) $x = 0.204081$, $y = -0.199958$, $r = 0.285714$, $\theta = -0.775193$; and iv) $x = 0.204081$, $y = 0.199958$, $r = 0.285714$, $\theta = 0.775193$.

In order to find $\frac{dy}{dx}$ the commands from the utility file DIF_APPS need to be loaded into the work area. We do this with the `Transfer Merge` command or `Transfer Load Utility` command. Enter $\boxed{\text{DIF\_APPS}}$ for the filename. To find $\frac{dy}{dx}$ for $r = 2\sqrt{\cos(2t)}$, Author

$$\boxed{\texttt{POLAR\_DIF(2sqrt(cos(2t)),t,1)}}.$$

Execute `Simplify` to obtain the following result:

```
50:   POLAR_DIF (2 √(COS (2 t)), t, 1)

          COS (t) COS (2 t) - SIN (t) SIN (2 t)
51:   - ─────────────────────────────────────────
          SIN (t) COS (2 t) + COS (t) SIN (2 t)
```

The derivative for $r = 1 - \cos t$ is determined the same way. Author

$$\boxed{\texttt{POLAR\_DIF(1-cost,t,1)}}.$$

Simplify to get $\frac{dy}{dx}$

```
52:   POLAR_DIF (1 - COS (t), t, 1)

                    2
          2 COS (t)  - COS (t) - 1
53:   - ──────────────────────────
            SIN (t) (2 COS (t) - 1)
```

110   Chapter 2. Examples

The formula for the curvature of a polar function is found with the POLAR_CURVATURE command. For $r = 2\sqrt{\cos(2t)}$, Author

$$\boxed{\texttt{POLAR\_CURVATURE(2sqrt(cos(2t)),t)}}.$$

Execute the Simplify command to produce the following result:

```
54:  POLAR_CURVATURE (2 √(COS (2 t)), t)

55:  - 1.5 √(COS (2 t)) SIGN (SIN (t) COS (2 t) + COS (t) SIN (2 t))
```

For $r = 1 - \cos t$, the curvature formula is found the same way. Author

$$\boxed{\texttt{POLAR\_CURVATURE(1-cost,t)}}.$$

Simplify this command to obtain after a long wait the following long and messy result that is truncated on the right:

```
56:  POLAR_CURVATURE (1 - COS (t), t)

                                                         3
            SIGN (SIN (t) (2 COS (t) - 1)) (2 COS (t)  + COS (t) (2 SIN
57:  -  ─────────────────────────────────────────────────────────────
                    4            3           2            2
            (4 COS (t)  - 4 COS (t)  + COS (t)  (4 SIN (t)  - 3) - 2 COS (t) (2 SIN
```

The result can be simplified by executing Manage Trigonometry and Collect toward Sines. Simplify to obtain

```
            3 √2 SIGN (SIN (2 t) - SIN (t)) (COS (3 t) - 2 COS (2 t) - COS (t) + 2)
58:  -  ─────────────────────────────────────────────────────────────────────────
                                                            3/2
                              8 (COS (2 t) - 1) (1 - COS (t))
```

Example 2.12 Building a Pool

It always was the biggest fish I caught that got away.

— Eugene Field [1890]

Subject: Finding the best dimensions for a pool by using optimization
References: Sections 1.7 and 1.9
Problem: We need to build an in-ground pool for a fish hatchery. It must contain 5300 ft^3 and have a linear-sloping bottom from 2 feet to 12 feet. The two ends will be rectangular. A sketch of the pool with the length labelled y and the width labelled x is as follows:

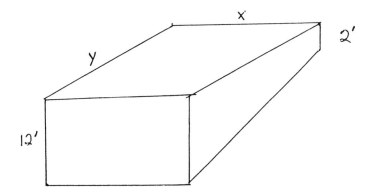

We need to determine the length y and the width x that results in the least building cost. The excavation costs $0.28 per ft³ for the entire volume of the pool. The material for the rectangular ends of the pool costs $0.70 per ft². The material for the sloped sides of the pool costs $1.25 per ft². The bottom is made of a heavier material that costs $2.95 per ft².

Solution: First, let's determine the equation for the volume of the pool. To do this, we use a simple formula of the average depth times the length times the width. So for this pool the volume is $7xy$ which must equal 5300.

We also need to determine the length of the sloped bottom which is simply the hypotenuse of a triangle with sides y and 10. Therefore, the length of the bottom is $\sqrt{y^2 + 100}$.

The cost function C is a sum of the evacuation cost $0.28(7xy)$, the material cost of the two ends $0.70(4x) + 0.70(10x)$, the material cost of the two sides $2(1.25)(7y)$, and the material cost of the bottom $2.95(x\sqrt{y^2 + 100})$.

Therefore, we want to find the minimum of the cost function

$$C = 0.28(7xy) + 0.7(14x) + 2.5(7y) + 2.95(x\sqrt{y^2 + 100})$$

when $7xy = 5300$. First Author $\boxed{\text{7xy=5300}}$ and soLve for y. The result is

```
1:    7 x y = 5300

              5300
2:    y =  ───────
               7 x
```

Then Author the cost function c

$\boxed{\text{0.28(7xy)+0.7(14x)+2.5(7y)+2.95(x sqrt(y^2+100))}}$.

112 Chapter 2. Examples

Then execute Manage Substitute and replace y with $5300/(7x)$. Simplify this expression to get the new cost function in terms of x only. The resulting Derive display with expression #4 truncated on the right is

$$3:\quad 0.28\,(7\,x\,y) + 0.7\,(14\,x) + 2.5\,(7\,y) + 2.95\,(x\,\sqrt{(y^2 + 100)})$$

$$4:\quad 0.28\left[7\,x\,\frac{5300}{7\,x}\right] + 0.7\,(14\,x) + 2.5\left[7\,\frac{5300}{7\,x}\right] + 2.95\left[x\,\sqrt{\left[\frac{5300}{7\,x}\right]^2 + 100}\right]$$

$$5:\quad \frac{59\,\sqrt{(49\,x^2 + 280900)}\,\text{SIGN}(x)}{14} + \frac{49\,x}{5} + \frac{13250}{x} + 1484$$

To find the minimum of this function, we take its derivative by issuing the Calculus Differentiate command with differentiation with respect to x and 1st order. Execute Simplify to perform the operation. The display shows this as

$$6:\quad \frac{d}{dx}\left[\frac{59\,\sqrt{(49\,x^2 + 280900)}\,\text{SIGN}(x)}{14} + \frac{49\,x}{5} + \frac{13250}{x} + 1484\right]$$

$$7:\quad \frac{413\,|x|}{2\,\sqrt{(49\,x^2 + 280900)}} - \frac{13250}{x^2} + \frac{49}{5}$$

Next, place Derive in the Approximate mode by using the Options Precision command. Then execute soLve with bounds of 0 and 40 to get the following critical point:

$$7:\quad \frac{413\,|x|}{2\,\sqrt{(49\,x^2 + 280900)}} - \frac{13250}{x^2} + \frac{49}{5}$$

$$8:\quad x = 26.1202$$

Before we forget, return to the Exact mode using the Options Precision command. Determine the related length y by using Manage Substitute to replace x with 26.1202 in the expression for y (expression #2). The result after executing approX is as follows:

9: $y = \dfrac{5300}{7\ 26.1202}$

10: $y = 28.9868$

Let's plot the cost function as a function of x. Highlight the expression for cost in terms of x (#5) and issue the Plot command. Set up the plot region by issuing Scale and setting x-scale to 8 and y-scale to 3000. Move the cross to $x = 15, y = 0$ and execute Center. Select the Plot command to obtain the following graph:

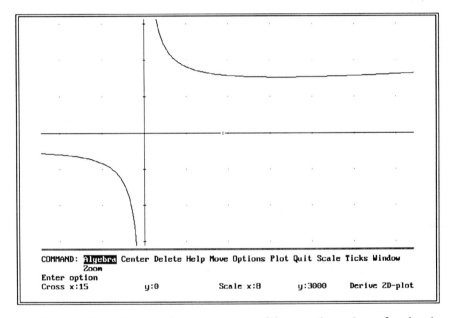

We see that the cost is not very sensitive to the value of x in the interval $20 < x < 40$. Replace x with 26.1202 in expression #5 using Manage Substitute and Simplify to obtain the total cost of the pool:

11: $\dfrac{59\ \sqrt{(49\ 26.1202^2 + 280900)\ \text{SIGN}\ (26.1202)}}{14} + \dfrac{49\ 26.1202}{5} + \dfrac{13250}{26.1202} + 14$

12: $\dfrac{59\ \sqrt{7858274438849}}{70000} + \dfrac{7337321538849}{3265025000}$

13: 4610

Example 2.13: Taylor Polynomial

...over the entrances to the gates of the temple of science are written the words: Ye must have faith.

—Max Planck [1932]

Subject: Taylor polynomial approximations
References: Sections 1.7, 1.9, and 1.13
Problem: Find the Taylor polynomial approximations of degrees 3, 4, 7, and 9 for $\sin x$ about $x = 0$. Then approximate the solution to the nonlinear differential equation

$$y' = \sin(xy) \text{ with } y(0) = 1$$

using 3rd, 4th, 7th, and 9th degree Taylor polynomials.

Solution: The Taylor polynomial approximation for a function can be obtained in two ways. The first way is to use the `Calculus Taylor` menu command and answer the queries for the variable, degree, and point. The second way is to use the in-line `TAYLOR` command which requires this information as arguments in the command.

In order to obtain numerical coefficients for the polynomial, place Derive in the `Approximate` mode using the `Options Precision` command. Using the second method for the third degree polynomial approximation, Author

$$\boxed{\text{TAYLOR(sinx,x,0,3)}}.$$

Then issue `Simplify` to obtain

```
1:    TAYLOR (SIN (x), x, 0, 3)
                           3
2:    x - 0.166666 x
```

Using the first method for the fourth degree polynomial approximation, we Author $\boxed{\text{sinx}}$. Issue the `Calculus Taylor` command. Respond to the input queries with Variable: x, Degree: 4, and Point: 0. Execute the `Simplify` command to get

```
3:    SIN (x)

4:    TAYLOR (SIN (x), x, 0, 4)
                           3
5:    x - 0.166666 x
```

We see that the 3rd and 4th degree polynomial approximations are the same because the coefficient of the x^4 is evaluated as 0. Produce the 7th and 9th degree polynomials using either of the two methods. The results of these operations are as follows:

```
6:     TAYLOR (SIN (x), x, 0, 7)

              -4  7                 5              3
7:   - 1.98412 10   x  + 0.00833333 x  - 0.166666 x  + x

8:     TAYLOR (SIN (x), x, 0, 9)

             -6  9          -4  7                 5              3
9:   2.75573 10   x - 1.98412 10   x  + 0.00833333 x  - 0.166666 x  + x
```

Let's plot these polynomials (degrees 3, 7, and 9) along with the function $\sin x$ to compare their accuracy. Assemble the four functions into a vector with appropriate components by *Authoring* [#3,#5,#7,#9]. Then issue the Plot Plot command. The graphs using the default plotting parameters are as follows:

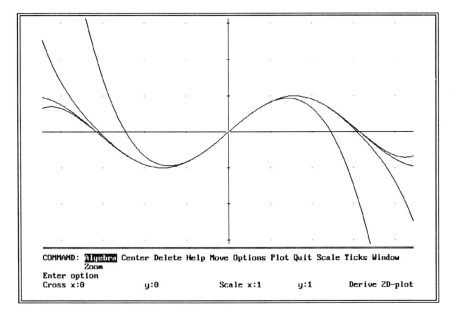

All of the polynomials seem to approximate $\sin x$ to the plotting accuracy in the interval $-1.25 < x < 1.25$ and the higher degree polynomials seem to improve the accuracy. To get a more global view of the functions, execute the Scale command and set the x-scale to 3. The larger-scaled plot is as follows:

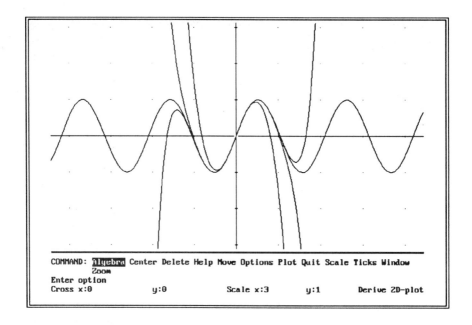

An approach to approximating solutions to nonlinear differential equations is to expand the functions into Taylor polynomials accurate in some neighborhood of the point of expansion. Derive has a command to find the nth-degree truncated Taylor-series solution in the utility file ODE_APPR. Once the file ODE_APPR is loaded into the work area with either the Transfer Load Utility or Transfer Merge command, the command to approximate the solution with a 3rd degree polynomial is

$$\boxed{\text{TAY\_ODE1}(\sin(xy),x,y,0,1,3)}.$$

Simplify this operation to obtain the following result

```
25:   TAY_ODE1 (SIN (x y), x, y, 0, 1, 3)
                2
26:   0.5 x  + 1
```

The higher degree approximations are determined by a similar procedure. Since only the last argument in the command needs to be changed, the highlight and F3 key can be helpful in reducing the retyping of the entire command. The higher degree approximations take considerable time. The results of the three operations (degrees 4, 7, and 9) are as follows:

```
27:   TAY_ODE1 (SIN (x y), x, y, 0, 1, 4)

                    4           2
28:   0.0833333 x  + 0.5 x  + 1

29:   TAY_ODE1 (SIN (x y), x, y, 0, 1, 7)

                    6                   4           2
30:  - 0.0263888 x  + 0.0833333 x  + 0.5 x  + 1

31:   TAY_ODE1 (SIN (x y), x, y, 0, 1, 9)

                    8                   6                   4           2
32:  - 0.0215525 x  - 0.0263888 x  + 0.0833333 x  + 0.5 x  + 1
```

Once again, the easiest way to plot these functions all at once is to place them in a vector as components through Author of the input

$$[\#26,\#28,\#30,\#32].$$

Then go to the Plot window and Scale the x-scale to 3 and the y-scale to 10. Now execute Plot. The plots of these functions reveal symmetry with respect to both axes and provide visualization of the solution behavior near the initial point $x = 0$.

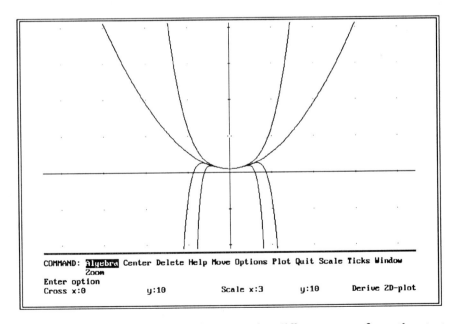

However, the solution trajectories are quite different away from the start point, which shows the poor nature of the approximations outside about $|x| < 1.2$. This technique is only good for analyzing local behavior.

Example 2.14 Separable Differential Equation

There is nothing so captivating as new knowledge.

— Peter Latham [1870]

Subject: Solving the logistics equation, a separable differential equation, using utility file ODE1
References: Sections 1.9 and 1.14
Problem: Find the solution to the modified logistics equation
$$\frac{dP}{dt} = P(a - bP)(1 - cP^{-1})$$
with $a, b, c > 0$ and $P(t) = 1000$ at $t = 0$.

Solution: This is a separable differential equation; therefore the function Separable in file ODE1 can be used to solve the equation. The utility file ODE1 is loaded using the Transfer Load Utility command and giving ODE1 for the file name.

If the differential equation is in the form $y' = p(x)q(x)$ with $y(x_0) = y_0$, then the form of the operation to solve the equation is:

$$\text{SEPARABLE}(p(x), q(y), x, y, x_0, y_0).$$

Before this command is entered for our example, the given conditions on the parameters $a, b,$ and c must be established. This is done with the Declare Variable command making $a, b,$ and c all Positive. This has to be done one variable at a time. The keystrokes **d v** result in the following menu display:

```
DECLARE VARIABLE name: _
Enter name or type "default"
User                         Free:100%        Derive Algebra
```

Then the sequence of screen displays showing the input to declare the variable a positive are as follows:

```
DECLARE VARIABLE name: a

Enter name or type "default"
User                              Free:100%          Derive Algebra
```

```
DECLARE VARIABLE: Domain Value

Enter option
User                              Free:100%          Derive Algebra
```

```
DECLARE VARIABLE DOMAIN: Positive Nonnegative Real Complex Interval

Select domain of a
User                              Free:100%          Derive Algebra
```

Now, to solve this equation, execute the Author command and enter:

$$\boxed{\text{SEPARABLE}(1,\ P(a-bP)(1-c/P),t,P,0,1000)}.$$

The command line and working area shows this as:

```
AUTHOR expression: separable(1,p(a-bp)(1-c/p),t,p,0,1000)_

Enter expression
User                              Free:98%           Derive Algebra
```

$$27:\ \text{SEPARABLE}\left[1,\ p\ (a-b\ p)\left[1-\frac{c}{p}\right],\ t,\ p,\ 0,\ 1000\right]$$

Simplify this expression to obtain the implicit solution:

$$28:\ \frac{\text{LN}\ (1000\ b-a)}{a-b\ c} - \frac{\text{LN}\ (b\ p-a)}{a-b\ c} - \frac{\text{LN}\ (1000-c)}{a-b\ c} + \frac{\text{LN}\ (p-c)}{a-b\ c} - t = 0$$

In order to obtain an explicit form for $P(t)$, ask Derive to soLve for P. Derive just returns the original implicit equation. This indicates it needs help. *Don't give up.* Help Derive by making it do what you would try. Clear the fractions and then exponentiate both sides of the equation. This can be done without retyping the terms in the equation by making use of the keystroking hints in Section 1.3. First let's multiply both sides of the equation by $(a - bc)$. Highlight the equation, select Author and press **F4** (or type expression #28). Then enter $\boxed{(\text{a-bc})}$ at the end of the command line. Press **Enter** and issue Simplify. The results are

$$30: \left[\frac{LN\,(p-c)}{a-bc} - \frac{LN\,(b\,p-a)}{a-bc} + \frac{LN\,(1000\,b-a)}{a-bc} - \frac{LN\,(1000-c)}{a-bc} - t = \right.$$

$$31: LN\,(1000\,b-a) - LN\,(b\,p-a) - LN\,(1000-c) + LN\,(p-c) - t\,(a-b\,c) =$$

The procedure and keystrokes to exponentiate are (i) highlight the left terms of the equation using the ← key, (ii) Author and enter $\boxed{\text{Alt-e\textasciicircum F4}}$, (iii) highlight the right-most term of the equation using the → key, and (iv) enter $\boxed{= \text{Alt-e\textasciicircum F4}}$. The working expression should be

$$32: \; \hat{e}^{\,LN\,(1000\,b-a) - LN\,(b\,p-a) - LN\,(1000-c) + LN\,(p-c) - t\,(a-b\,c)}$$

Now soLve for P to get

$$33: \; p = \frac{a\,\hat{e}^{a\,t}\,(c-1000) - c\,\hat{e}^{b\,c\,t}\,(a-1000\,b)}{b\,\hat{e}^{a\,t}\,(c-1000) + \hat{e}^{b\,c\,t}\,(1000\,b-a)}$$

To check the initial condition highlight the expression and execute Manage Substitute. Enter the value 0 for the variable t. Just press **Enter** when queried for replacement for the other unknowns. The result is

$$35:\quad p = \frac{a\,\hat{e}^{a\,0}(c - 1000) - c\,\hat{e}^{bc\,0}(a - 1000\,b)}{b\,\hat{e}^{a\,0}(c - 1000) + \hat{e}^{bc\,0}(1000\,b - a)}$$

Now, Simplify this expression to achieve the given initial value of 1000. To check the solution, Author and enter the operator shown in the following input line.

```
AUTHOR expression: dif(p,t)-p(a-bp)(1-c/p)_

Enter expression
Simp(35)              D:ODE1.MTH              Free:86%
```

Then, Manage-Substitute the expression for the solution P (for this example, this is the right-hand side of expression #33) into the operator using the highlight and F3 key. Simplify to get the following result:

$$38:\quad \frac{d}{dt}\,\frac{a\,\hat{e}^{a\,t}(c - 1000) - c\,\hat{e}^{bc\,t}(a - 1000\,b)}{b\,\hat{e}^{a\,t}(c - 1000) + \hat{e}^{bc\,t}(1000\,b - a)} - \frac{a\,\hat{e}^{a\,t}(c - 1000) - c\,\hat{e}^{b}}{b\,\hat{e}^{a\,t}(c - 1000) + \hat{e}^{bc}}$$

$$39:\quad 0$$

which verifies the result.

Let's use some specific values for $a, b,$ and c so we can plot one of the solution curves. We choose $a = 1$, $b = 0.5$, and $c = 0.5$. Manage Substitute these values into the solution (expression #33) and Simplify to get

$$40: \quad p = \frac{1 \hat{e}^{1\,t}(0.5 - 1000) - 0.5 \hat{e}^{0.5\,0.5\,t}(1 - 1000 \cdot 0.5)}{0.5 \hat{e}^{1\,t}(0.5 - 1000) + \hat{e}^{0.5\,0.5\,t}(1000 \cdot 0.5 - 1)}$$

$$41: \quad p = \frac{2\,(1999\,\hat{e}^{3t/4} - 499)}{1999\,\hat{e}^{3t/4} - 1996}$$

Plot Plot this solution with the default scale to show the solution graphically with these parameter values as:

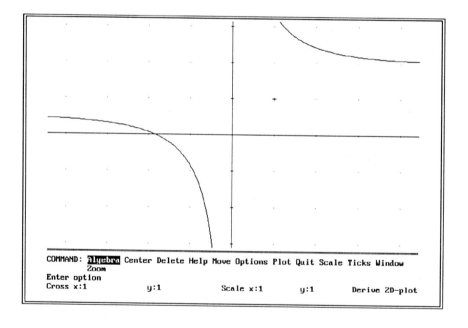

This plot does not show much of the overall behavior of the solution. This is because the default plotting window shows approximately $-4.5 < x < 4.5$ and $-3 < y < 3$. Let's change the window size by executing Scale and setting x-scale $= 1$ and y-scale $= 100$. The new plot is automatically redrawn to the new scale as follows:

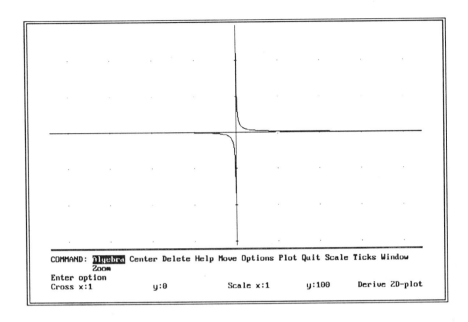

Example 2.15 Exact Differential Equation

A collection of facts is no more a science than a heap of stones is a house.

—Jules Henri Poincare [1903]

Subject: Solving an exact differential equation using utility file ODE1
References: Sections 1.9 and 1.14
Problem: Find the particular solution of
$(y^2 \cos x - 3x^2 y - e^x - 1) + (2y \sin x - x^3 + \ln y + 2)y' = 0$, with $y = 1$ at $x = 0$.

Solution: Exact differential equations are solved by integration. The command EXACT_TEST in file ODE1 can be used to check for exactness, and the command EXACT is used to solve the equation. The utility file ODE1 is loaded using the Transfer Load Utility command and giving ODE1 for the file name.

The differential equation is already in the proper form of $p(x, y) + q(x, y)y' = 0$, therefore the form of the exactness check is

$$\text{EXACT\_TEST}(p(x, y), q(x, y), x, y).$$

124   Chapter 2. Examples

The equation is exact if the *Simplified* result is equivalent to 0. The input command for the exactness check for this problem is

EXACT_TEST(y^2cosx-3x^2y-Alt-e^x-1,2y sinx-x^3+lny+2,x,y).

This displays in the work area as

28:   EXACT_TEST (y^2 COS (x) - 3 x^2 y - ê^x - 1, 2 y SIN (x) - x^3 + LN (y) + 2, x,

Simplify to obtain 0, which indicates the equation is exact.

In order to solve an exact equation, the command EXACT is used. The argument of the command are the same as those of EXACT_TEST along with the initial condition of the equation. For this problem, the command is entered as:

AUTHOR expression: EXACT(y^2COS(x)-3x^2y-ê^x-1,2ySIN(x)-x^3+LN(y)+2,x,y,0,1)
Enter expression
User   Free:94%   Derive Algebra

Most of this expression can be entered without retyping by using the highlight and **F3** key to obtain parts of the previous expression (see Section 1.3). Simplify this expression to obtain the solution. The command and its result are shown below.

29:   EXACT (y^2 COS (x) - 3 x^2 y - ê^x - 1, 2 y SIN (x) - x^3 + LN (y) + 2, x, y, 0

30:   - ê^x + y LN (y) + y^2 SIN (x) - x^3 y - x + y = 0

We now try to solve explicitly for *y* using the soLve command. However, the result shows there is a problem. The unchanged expression implies that Derive can not help. This implicit equation is transcendental in *y*, so it is impossible to determine an explicit solution for *y* in terms of built-in functions. This equation is also transcendental in *x*, so Derive can't solve for *x* in terms of *y* either. This is unfortunate since Derive version 2 does not plot implicit functions. However, don't despair, we do have the implicit solution to the differential equation because of Derive's help.

In fact, we may be able to plot a few explicit solution points. First set the Options Precision command to Approximate. Then use the

Manage Substitute command to replace x with specific values. Then execute the soLve command to find y via the bisection method. The search bounds for the method may have to be adjusted for different values of x. The results of these steps for $x = 1$, 2, and 3 are as follows:

$$31: -\hat{e}^1 + y \operatorname{LN}(y) + y^2 \operatorname{SIN}(1) - 1^3 y - 1 + y = 0$$

$$32: y = 1.78563$$

$$33: -\hat{e}^2 + y \operatorname{LN}(y) + y^2 \operatorname{SIN}(2) - 2^3 y - 2 + y = 0$$

$$34: y = 7.02438$$

$$35: -\hat{e}^3 + y \operatorname{LN}(y) + y^2 \operatorname{SIN}(3) - 3^3 y - 3 + y = 0$$

$$36: y = 149.834$$

These points, along with the point of the initial condition, can be placed in an array and plotted. Author

$$[[0,1], [1,1.785], [2,7.024], [3,149.834]].$$

Then select the Plot menu. Change the Scale to $x = 1$, $y = 50$ and execute Plot. The resulting plot is

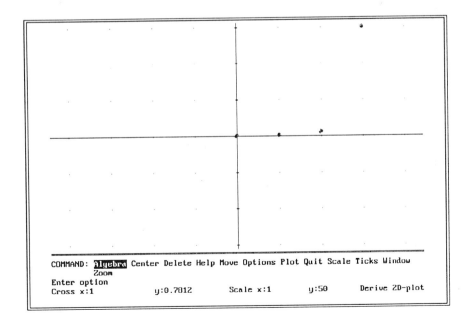

There are other ways to produce solution points for this differential equation. One possibility would be to use a numerical procedure to approximate the solution. See Section 2.18 for the use of numerical procedures to solve differential equations.

Example 2.16 Drug Doses

Eventually there will be, I hope, some people who will find it profitable to decipher this mess.

—Evariste Galois [1832]

Subject: Determining drug doses using a differential equation
References: Sections 1.9 and 1.14
Problem: The decrease in concentration of drugs in the bloodstream is proportional to the concentration of the drug. If the concentration in milligrams per milliliter of a drug at time t hours is represented by $D(t)$, then the differential equation model is $D'(t) = -kD(t)$. For the specific drug being administered, we know that $k = 0.27$, the minimum effective concentration is 2.2 mg/ml, and the maximum safe concentration is 3.4 mg/ml. It is reasonable to assume that the blood absorbs the drug instantaneously.

Devise a schedule of safe and effective doses of this drug for a body of a small animal with 1000 ml of blood.

Solution: Let's start with the dose at $t = 0$. We want to administer the maximum safe dosage and let it decrease in concentration until it reaches the minimum safe concentration. Then we will administer enough drug to raise the concentration back up to the maximum.

Therefore, at $t = 0$ we will administer $(3.4)(1000) = 3400$ mg of the drug to obtain a concentration of $D(0) = 3.4$. Next we need to find the time t when $D(t)$ decreases to 2.2. We will use the solution operators for differential equations in utility file ODE1.MTH. Load the functions from this file into the work area by executing Transfer Load Utility. Type ODE1 for the file name.

By writing the equation and initial conditions as $D'(t) + 0.27D(t) = 0$ and $D(0) = 3.4$, we see the equation is in linear form. Linear equations of the form

$$y' + p(x)y = q(x) \quad \text{with} \quad y = y_0 \text{ at } x = x_0,$$

are solved with the command

$$\boxed{\text{LINEAR1}(p,q,x,y,x_0,y_0)}.$$

Set up the solution operator for our model equation through the Author of

$$\boxed{\text{LINEAR1}(0.27,0,t,d,0,3.4)}.$$

Simplify this expression to obtain the solution shown in the following display:

```
27:   LINEAR1 (0.27, 0, t, d, 0, 3.4)

              - 27 t / 100
         17 ê
28:  d = ─────────────────
                5
```

We need to find the time t when $D(t) = 3.4e^{-0.27t} = 2.2$. To do this, Author

$$\boxed{3.4 \text{ Alt-e}\char`^(-2.7t)-2.2}$$

and execute soLve. Unfortunately, Derive will not solve for t in the Exact mode. We can approximate the solution for t using the bisection method by changing the mode by issuing Options Precision Approximate. The reexecute soLve with bounds of 0 and 10 to obtain the solution

```
              - 0.27 t
29:   3.4 ê            - 2.2

30:   t = 1.61228
```

By our stated plan, at $t = 1.6122$ hours, we administer $3,400 - 2,200$ or 1200 mg of the drug to increase the drug to its maximum safe concentration. We now know our drug dose schedule — administer 1200 mg of the drug every 1.6 hours.

Derive has several ways to represent this schedule by a discontinuous function. We could probably use the Step function or program the function using the IF command. Probably the easiest way to represent this function is through the use of the MOD function found in the MISC utility file. The MOD(a,b) command evaluates a modulo b or the remainder of a/b. This is just what is needed since the drug schedule repeats itself every 1.6122 hours so we only need to know how long since the last drug dose. In order to plot the cycles of this schedule. Author

$$\boxed{3.4 \text{ Alt-e}\char`^(-0.27 \text{ MOD}(t,1.6122))}.$$

Then select the `Plot` menu. Move the cross to $x = 4$ and $y = 2$ and execute `Center`. Leave the x- and y-scales at 1. Execute `Plot` to obtain the following plot of the drug concentration (mg/ml) over time.

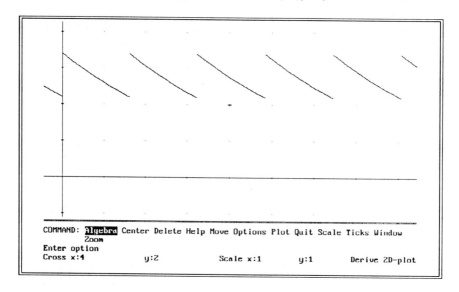

Notice how the concentration stays in the safe but effective region for all time.

Example 2.17 Electrical Circuit

...the aim of exact science is to reduce the problems of nature to the determination of quantities by operations with numbers.

— James Maxwell [1856]

Subject: The series electrical circuit
References: Sections 1.9 and 1.14
Problem: The basic, simple-loop electrical circuit consists of several parts: a time-dependent voltage source $E(t)$, an inductor L, a resistor R, and a capacitor C, arranged in series. Variations of this circuit containing more than one inductor, resistor, and capacitor are found throughout

homes. In these simple circuits, L, R, and C represent the loop's total induction, resistance, and capacitance, respectively. The problem is to insure that the electrical current in any loop does not exceed the safe wire capacity which is protected by the limit of the fuse or circuit breaker.

Solution: Of course, we could take the experimental approach of plugging everything into the circuit and seeing if the fuse blows. However, we choose a more analytical approach. The best way to determine the current is to set up a differential model using Kirchoff's laws. By letting $q(t)$ be the time-dependent charge on the capacitor, the model is

$$L\frac{d^2q}{dt^2} + R\frac{dq}{dt} + \frac{1}{C}q = E(t)$$

where L, R, and C are assumed to be constant for this circuit.

The applied voltage for a typical USA household alternating current can be modeled as

$$E(t) = 110\cos 2t \text{ volts.}$$

The Derive plot of this voltage is made with the Author command and by entering this expression $\boxed{\texttt{110 cos(2t)}}$ into the work area and using the Plot command. Good parameters for the Scale command are x-scale $= 5$ and y-scale $= 50$. Move the cross to $x = 20$ and $y = 0$ and execute the Center command. Don't forget to use the Tab key in order to move the cursor between different parts of the menu. After these parameters are set, then the Plot command is issued to begin drawing the graph of the function. The resulting voltage plot is shown.

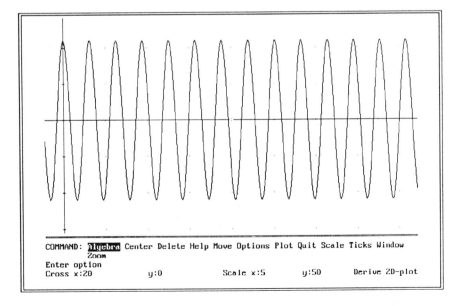

For a circuit with $L = 10$ henrys, $R = 50$ ohms, $C = 1/60$ farads, no initial charge on the capacitor ($q(0) = 0$), no initial current ($q'(0) = 0$), and $E(t)$ applied at $t = 0$ (q measured in volts, t measured in seconds), the equation to be solved is

$$10\frac{d^2q}{dt^2} + 50\frac{dq}{dt} + 60q = 110\cos 2t,$$

with $q(0) = 0$ and $q'(0) = 0$.

To solve this equation with Derive, first use the Transfer Merge command or the Transfer Load Utility command to obtain the utility file ODE2.MTH which has commands available to solve second-order, constant coefficient, nonhomogeneous equations in the form $y'' + ay' + by = r$ (Note: a and b have been substituted for the p and q in file ODE2 because q has already been used as the dependent variable in the equation). Next, place the equation in this standard form

$$\frac{d^2q}{dt^2} + 5\frac{dq}{dt} + 6q = 11\cos 2t$$

to identify $a = 5$ and $b = 6$.

Use the in-line command

$$\boxed{\text{LIN2\_TEST(5,6,t)}}$$

and Simplify to obtain the following result for the evaluation of the classification function.

```
25:   LIN2_TEST (5, 6, t)

26:   0.25
```

The positive value of 1/4 resulting from this command determines the next command in the solution sequence. Therefore, for this case, the command $\boxed{\text{LIN2\_POS(5,6,11cos(2t),t)}}$ computes the general solution to the equation.

For this problem, the command and resulting display after Simplify are as follows.

```
27:   LIN2_POS (5, 6, 11 COS (2 t), t)

              -2 t       -3 t    11 COS (2 t)    55 SIN (2 t)
28:   c1 ê       - c2 ê       + ─────────────  + ─────────────
                                      52              52
```

The general solution is

$$q(t) = c_1 e^{-2t} - c_2 e^{-3t} + \frac{11 \cos 2t}{52} + \frac{55 \sin 2t}{52}$$

using c_1 and c_2 for the arbitrary constants. To use c1 and c2 as the arbitrary constants in Derive, change the input mode to Word by using the Options Input command. *Note the minus sign before the c_2.*

The two initial conditions are imposed to find values for c_1 and c_2 using the command

```
IMPOSE_IV2(t, c1Alt-e^(-2t)-c2Alt-e^(-3t)+11cos(2t)/52+ 55sin(2t)/52,0,0,0,c1,c2)
```

The most efficient way to enter this is by *Authoring*

```
IMPOSE_IV2(t,#28,0,0,0,c1,c2)
```

Another method is to move the highlight to parts of previous expressions and using the **F3** key to bring these highlighted expressions into the work area. Simplify this expression to get the following values for c_1 and c_2 in the resulting vector.

```
29: IMPOSE_IV2 [t, c1 ê^(-2t) - c2 ê^(-3t) + 11 COS (2 t)/52 + 55 SIN (2 t)/52, 0,

30: [ c1 = - 11/4   c2 = - 33/13 ]
```

The solution to the differential equation can now be written as

$$q(t) = -\frac{11}{4}e^{-2t} + \frac{33}{13}e^{-3t} + \frac{11 \cos 2t}{52} + \frac{55 \sin 2t}{52}.$$

This is entered into Derive through the use of the Manage Substitute command on the expression for the general solution (expression #28). Execution of this command results in queries for values for each of the remaining variable in the expression. In this case just press **Enter** for the value for t (indicating no substitution) and the values of $-11/4$ and $33/13$ for values of c1 and c2, respectively. The resulting Derive display after executing Simplify for this is as follows.

```
31: [- 11/4] ê^(-2t) - [- 33/13] ê^(-3t) + 11 COS (2 t)/52 + 55 SIN (2 t)/52

32: - 11 ê^(-2t)/4 + 33 ê^(-3t)/13 + 11 COS (2 t)/52 + 55 SIN (2 t)/52
```

Before plotting this solution, it should be checked. Derive can help do this. Author the operator for the equation by entering

```
AUTHOR expression: 10 dif(q,t,2)+50 dif(q,t)+60q-110cos(2t)_

Enter expression
Sub(28)              D:ODE2.MTH              Free:100%
```

Use the Manage Substitute command to substitute the expression for the solution into the equation for the variable q. To do this, use the highlight and **F3** key to put the equation in the input line of the menu or use the # expression reference number (in this case #32). The display for this resulting command is truncated on the right in the following figure because of its length.

$$34: \quad 10 \left[\frac{d}{dt}\right]^2 \left[-\frac{11 \hat{e}^{-2t}}{4} + \frac{33 \hat{e}^{-3t}}{13} + \frac{11 \cos(2t)}{52} + \frac{55 \sin(2t)}{52} \right] +$$

Use the Simplify command to get verification that the function does, in fact, evaluate to 0. A check of the initial condition $q(0) = 0$ is made by substituting 0 for t and evaluating the expression.

Since the current $i(t)$ is equal to $q'(t)$, use Derive to get an expression for $i(t)$. One way to do this is to type the expression $\boxed{i = \text{DIF}(\#n,t)}$, where n is the statement number of the expression for $q(t)$ (#32 for this example) or instead of #n just highlight the expression for $q(t)$ and move it into the appropriate place in the work area with the **F3** key. The function displays as

$$36: \quad i = \frac{d}{dt}\left[\left[-\frac{11}{4}\right]\hat{e}^{-2t} - \left[\frac{33}{13}\right]\hat{e}^{-3t} + \frac{11 \cos(2t)}{52} + \frac{55 \sin(2t)}{52}\right]$$

Simplify this to obtain the following expression for $i(t)$.

$$37: \quad i = \frac{11 \hat{e}^{-2t}}{2} - \frac{99 \hat{e}^{-3t}}{13} + \frac{55 \cos(2t)}{26} - \frac{11 \sin(2t)}{26}$$

Probably, the easiest way to check that this function for the current does not exceed a certain amperage rating is to plot this function for $t > 0$ with a proper scale to see the global behavior. Highlight the expression for $i(t)$ and issue the Plot command. Use the Delete-All command, if there are previous plots being drawn. Set x-scale = 5 and y-scale = 2 with the Scale menu command. Move the cross to $x = 20$ and $y = 0$ and execute the Center command. Plot the function to obtain the following plot screen.

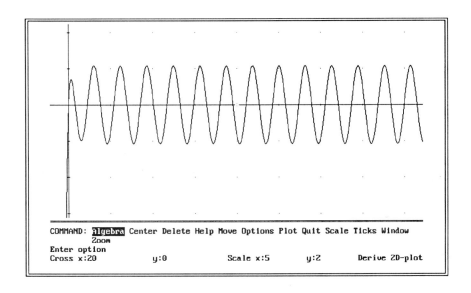

Move the cross using the direction arrows to determine that the maximum absolute current does not exceed a value of 2.3. Therefore, a 5 amp capacity (wiring and fuse) probably would be sufficient for this circuit.

While analyzing this solution, let's take note of the form of this long-term (steady-state) behavior. This behavior is more similar in form to the forcing function than the initial conditions. The initial conditions seem to affect only the short-term behavior. The graphing capabilities of Derive are very useful in making qualitative observations such as these. This can be explained by looking at the first two summands in #32. Both contain negative exponentials and therefore, die off as t increases. Therefore, the remaining trigonometric terms eventually dominate the solution.

Example 2.18 Numerical Solution of Differential Equations

When you have eliminated the impossible, whatever remains, however improbable, must be the truth..

— Sir Arthur Conan Doyle, *The Sign of Four* [1890]

Subject: Numerical solution of differential equations
References: Sections 1.9 and 1.14
Problem: Given the following first-order, nonlinear differential equation

$$y' = \frac{y^2 \sin(2x)}{x^2}, \quad \text{with} \quad y(1) = 2,$$

use the Euler and Runge-Kutta numerical methods to approximate solutions to the given equation with step-size $h = 0.15$ and number of steps $n = 50$.

Solution: The commands for these two numerical procedures are located in utility file ODE_APPR.MTH. Load this file through execution of `Transfer Load Utility` and entering `ODE_APPR`.

The command to perform the Euler method for a differential equation in the form $y' = r(x, y)$ is

$$\boxed{\texttt{EULER(r(x,y),x,y,}x_0\texttt{,}y_0\texttt{,h,n)}}.$$

Therefore, for this problem the Euler method is set up by the Author of the command

$$\boxed{\texttt{EULER(y\^{}2 sin(2x)/x\^{}2,x,y,1,2,0.15,50)}}.$$

The iteration is performed by executing `approX`. This command uses the Derive `ITERATE` command and, therefore, takes time and memory to execute. The result is a set of points (x, y), which approximate points on the solution curve. The command and the beginning of the set of points (truncated because of length) are as follows:

```
              ⎡  y² SIN (2 x)                        ⎤
15:   EULER   ⎢  ────────────, x, y, 1, 2, 0.15, 50  ⎥
              ⎣       x²                             ⎦

16:   [[1, 2], [1.15, 2.54557], [1.3, 3.09364], [1.45, 3.53154], [1.6, 3.74443],
```

Before we go to the `Plot` menu to plot these points, we check the range of the y values of the points. In order to see all 50 points, the `Ctrl-→` keys move the visible area to the right. A quick look shows the range of y values is $2 \le y \le 4$. Since we already know that $0 \le x \le 8.5$, we know the plot region to produce using rescaling and centering. First we select

Plot Scale and set the *x*-scale to 1.5 and *y*-scale to 1. Move the cross to $x = 4.5$ and $y = 2$ and execute Center. Then execute Plot to get the following display:

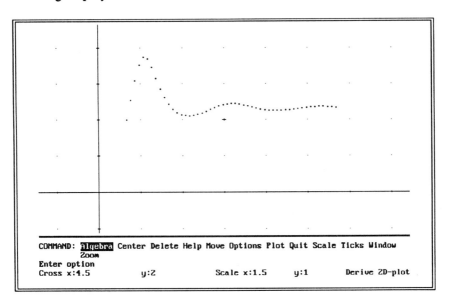

The Runge-Kutta method has a higher order of accuracy than the Euler method. Of course the cost of this accuracy is more computational work per step. Let's see if this method improves the solution to this problem. The command to perform the Runge-Kutta method is slightly different than the command for the Euler method. This is because the Runge-Kutta command is designed for systems of equations. For a scalar equation of the form $y' = r(x, y)$, the command is

$$\boxed{\text{RK}([r(x,y)], [x,y], [x_0, y_0], h, n)}$$

Therefore, for this problem the Runge-Kutta method is set up with the Author command and entering

$$\boxed{\text{RK}([(y^2 \sin(2x))/x^2], [x,y], [1,2], 0.15, 50)}$$

Execute the approX command to obtain a set of solution values. The command and its truncated result are as follows:

$$17: \quad \text{RK}\left[\left[\frac{y^2 \text{ SIN } (2\ x)}{x^2}\right], [x, y], [1, 2], 0.15, 50\right]$$

18: [[1, 2], [1.15, 2.55904], [1.3, 3.06231], [1.45, 3.37962], [1.6, 3.45205],

Execute `Plot Plot` to see these values produced with the Runge-Kutta method plotted on the same axes as the values determined using the Euler method.

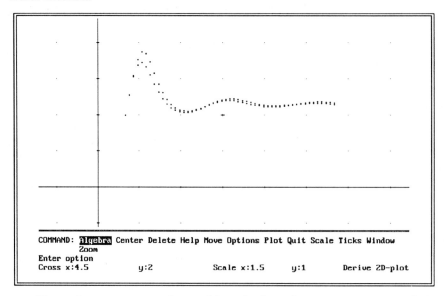

The points are pretty close, although there is more error toward the beginning (left side) of the interval and an overshoot of 10% may exceed the allowable tolerance.

What happens if we reduce the step size to 0.1 and compute for 75 steps. The result of plotting these 75 points on the same axes as the other two plots is as shown:

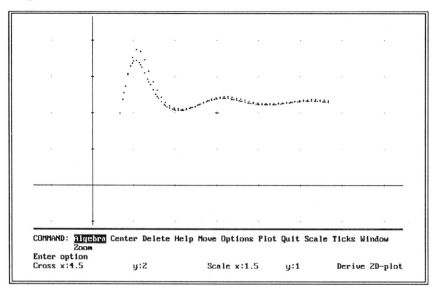

Example 2.19 Linear Algebra

Mathematics is the science which draws necessary conclusions.

—Benjamin Peirce, *Linear Associative Algebra* [1870]

Subject: Solving systems of linear algebraic equations
Reference: Section 1.8
Problem: In problem solving, there are times when a solution to a linear system of algebraic equations is needed. There are several ways to solve systems of linear equations using Derive. The following examples demonstrate three of the ways along with a few subtle extras.
Solution: For the first example, let's find the solution to the 4×4 linear system

$$2x + y + 2z - 3w = 0$$
$$4x + y + z + w = 15$$
$$6x - y - z - w = 5$$
$$4x - 2y + 3z - w = 2$$

One way to solve this is to Author the four equations as components of a vector in Derive's work area. The expression is entered and displayed (truncated) as

```
AUTHOR expression: [2x+y+2z-3w=0,4x+y+z+w=15,6x-y-z-w=5,4x-2y+3z-w=2]_

Enter expression
                              Free:100%            Derive Algebra
```

```
1:   [2 x + y + 2 z - 3 w = 0, 4 x + y + z + w = 15, 6 x - y - z - w = 5, 4 x -
```

Then, soLve this expression to obtain the solution

```
2:   [x = 2, y = 3, z = 1, w = 3]
```

The answer is produced, very conveniently, in less than one second.

Let's solve the same system using matrices with Derive. If the above system is written as a matrix-vector system $\mathbf{A}\vec{x} = \vec{f}$, then the solution vector $\vec{x}$ can be determined by $\mathbf{A}^{-1}\vec{f}$. For this problem the solution is found by the evaluation of

$$\begin{bmatrix} 2 & 1 & 2 & -3 \\ 4 & 1 & 1 & 1 \\ 6 & -1 & -1 & -1 \\ 4 & -2 & 3 & -1 \end{bmatrix}^{-1} \begin{bmatrix} 0 \\ 15 \\ 5 \\ 2 \end{bmatrix}.$$

This expression is entered into the working area through the Author command by typing

```
[[2,1,2,-3],[4,1,1,1],[6,-1,-1,-1], [4,-2,3,-1]]^(-1).[0,15,5,2]
```

Don't forget the dot (.) between the matrix and the vector. Derive uses the dot as the matrix multiplication symbol. Derive displays this expression as

3: $\begin{bmatrix} 2 & 1 & 2 & -3 \\ 4 & 1 & 1 & 1 \\ 6 & -1 & -1 & -1 \\ 4 & -2 & 3 & -1 \end{bmatrix}^{-1} \cdot [0, 15, 5, 2]$

and the Simplify command gives the answer

4: [2, 3, 1, 3]

These nice integer answers are typical of textbook problems, but such simple results are not expected very often in realistic problems. Not only that, but also solutions to linear systems with integer coefficients can result in quite messy real numbers. For instance, let's change the last equation in the above system to $2x - 2y + 3z - w = 2/7$. To do this easily, highlight the vector of equations previously entered and use the Author command and F3 key to bring that expression into the working area. The Ctrl-S and Ctrl-D keys provide movement of the cursor in the expression so the coefficients in the last expression can be changed. This is an instance when smart keystroking saves a lot of time. No one wants to retype all 4 of these equations. Enter this expression and soLve to obtain the solution

6: $[2x + y + 2z - 3w = 0, 4x + y + z + w = 15, 6x - y - z - w = 5, 2x -$

7: $[x = 2, y = \dfrac{361}{147}, z = \dfrac{211}{147}, w = \dfrac{457}{147}]$

in only a second or so depending on the speed of the computer being used. These rational numbers can be converted to approximate 6-digit decimal equivalents with the `approX` command to get

$$7: \quad \left[x = 2, \; y = \frac{361}{147}, \; z = \frac{211}{147}, \; w = \frac{457}{147}\right]$$

$$8: \quad [x = 2, \; y = 2.45578, \; z = 1.43537, \; w = 3.10884]$$

Derive has no problem handling parameters in the equations. For instance, let's change the 2/7 in the last equation to the parameter α (it's time to see Derive do Greek letters). Follow the keyboarding steps above to obtain the previous set of equations and change the 2/7 to α. To get Derive to use the Greek letter α, type **ALT-a**. soLve this expression with the solve variables set to $x, y, z,$ and w to obtain

$$10: \quad [2x + y + 2z - 3w = 0, \; 4x + y + z + w = 15, \; 6x - y - z - w = 5, \; 2x -$$

$$11: \quad \left[x = 2, \; y = -\frac{5\alpha - 53}{21}, \; z = \frac{4\alpha + 29}{21}, \; w = \frac{\alpha + 65}{21}\right]$$

Now we will do another example. This example shows when and how matrix manipulation can be used effectively to solve systems of equations. Sometimes the problem requires solutions for several systems with different right-hand sides. The following matrix equation can represent 4 different 3×3 systems of equations. In matrix form, this example problem is

$$\begin{bmatrix} 1 & -1 & 1 \\ 1 & -2 & -2 \\ 2 & 1 & 3 \end{bmatrix} \begin{bmatrix} x & u & m & r \\ y & v & n & s \\ z & w & p & t \end{bmatrix} = \begin{bmatrix} 3 & 1 & -6 & 3/2 \\ 0 & -1 & 1 & 0 \\ 4 & 3 & -9 & 11/2 \end{bmatrix}.$$

Another way to enter matrices into Derive is through the `Declare Matrix` command, which prompts the user for matrix size (rows and columns) and the elements of the matrix. Let's enter the two matrices given above containing numbers as separate working expressions in Derive using the `Declare Matrix` command. The first matrix has 3 rows and 3 columns. The second matrix has 3 rows and 4 columns. The result of the input of the elements in these matrices is the following screen display of the two matrices.

$$12: \begin{bmatrix} 1 & -1 & 1 \\ 1 & -2 & -2 \\ 2 & 1 & 3 \end{bmatrix}$$

$$13: \begin{bmatrix} 3 & 1 & -6 & \dfrac{3}{2} \\ 0 & -1 & 1 & 0 \\ 4 & 3 & -9 & \dfrac{11}{2} \end{bmatrix}$$

The solution matrix is obtained by multiplying the inverse of the 3x3 matrix times the 3x4 matrix. This can be accomplished with by Author of the command #n^(-1).#k, where n and k are the numbers of the expression for these matrices in the work area. Don't forget the dot between the matrices. For the above Derive screen, the command is #12^(-1).#13. This displays as

$$14: \begin{bmatrix} 1 & -1 & 1 \\ 1 & -2 & -2 \\ 2 & 1 & 3 \end{bmatrix}^{-1} \cdot \begin{bmatrix} 3 & 1 & -6 & \dfrac{3}{2} \\ 0 & -1 & 1 & 0 \\ 4 & 3 & -9 & \dfrac{11}{2} \end{bmatrix}$$

Simplify this expression to obtain the solution matrix

$$15: \begin{bmatrix} \dfrac{1}{2} & \dfrac{1}{2} & -1 & 2 \\ -\dfrac{9}{8} & \dfrac{1}{8} & 2 & \dfrac{3}{4} \\ \dfrac{11}{8} & \dfrac{5}{8} & -3 & \dfrac{1}{4} \end{bmatrix}$$

There is one other matrix operation to discuss, row reduction. The Derive command to perform this operation is ROW_REDUCE. Let's use this command to solve the matrix equation,

$$\begin{bmatrix} 4 & 1 & 4 & -3 \\ 4 & -1 & 1 & -1 \\ 4 & -2 & -1 & -1 \\ 2 & -2 & 0 & -1 \end{bmatrix} \begin{bmatrix} w \\ x \\ y \\ z \end{bmatrix} = \begin{bmatrix} 1 \\ 11 \\ -5 \\ -12 \end{bmatrix}.$$

To do this, enter the augmented matrix,

$$\begin{bmatrix} 4 & 1 & 4 & -3 & 1 \\ 4 & -1 & 1 & -1 & 11 \\ 4 & -2 & -1 & -1 & -5 \\ 2 & -2 & 0 & -1 & 12 \end{bmatrix}$$

into the work area using one of the two methods mentioned previously. Then issue the command ROW_REDUCE(#n), where n is the expression number that contains the augmented matrix. The display showing the result is as follows:

```
         ⎡ 4   1   4  -3   1 ⎤
         ⎢ 4  -1   1  -1  11 ⎥
16:      ⎢ 4  -2  -1  -1  -5 ⎥
         ⎣ 2  -2   0  -1  12 ⎦

                      ⎡ 4   1   4  -3   1 ⎤
                      ⎢ 4  -1   1  -1  11 ⎥
17:  ROW_REDUCE       ⎢ 4  -2  -1  -1  -5 ⎥
                      ⎣ 2  -2   0  -1  12 ⎦
```

Execute Simplify to perform the row reduction. The resulting matrix is shown below.

18: $\begin{bmatrix} 1 & 0 & 0 & 0 & -\frac{23}{22} \\ 0 & 1 & 0 & 0 & -\frac{152}{11} \\ 0 & 0 & 1 & 0 & \frac{164}{11} \\ 0 & 0 & 0 & 1 & \frac{149}{11} \end{bmatrix}$

Now the solution vector can be directly determined as

$$\begin{bmatrix} \frac{-23}{22} \\ \frac{-152}{11} \\ \frac{164}{11} \\ \frac{149}{11} \end{bmatrix}$$

Example 2.20 Series

The eternal silence of these infinite places terrifies me.

— Blaise Pascal [1670]

Subject: Summations and series
Reference: Section 1.7
Problem: Evaluate the following summations and series:

$$S_1 = \sum_{k=1}^{10} \frac{1}{k}$$

$$S_2 = \sum_{k=1}^{100} \frac{1}{k}$$

$$S_3(n) = \sum_{k=1}^{n} (2k - 1)$$

$$S_4 = \sum_{k=0}^{\infty} 4(0.8)^k$$

$$S_5 = \sum_{k=1}^{\infty} \frac{1}{k(k+1)}$$

$$S_6 = \sum_{k=1}^{\infty} \frac{2}{10^k}$$

Solution: Derive has two ways to set up summations. The first way uses the Calculus Sum menu command and provide answers to the queries of the submenu. The second way uses the SUM in-line command and provide the appropriate arguments for the function.

Using the first method for S_1, Author the expression $\boxed{1/k}$. Then, select Calculus Sum and enter k for the summation variable, 1 for the lower limit, and 10 for the upper limit. Simplify to obtain:

```
1:      1
        ─
        k

2:     10   1
       Σ   ─
       k=1  k

3:    7381
      ────
      2520
```

The Derive in-line command to set up the summation for S_2 is
$\boxed{\text{SUM}(1/k,k,1,100)}$.
Author this command and approX this time to see a 6-digit decimal approximation. The results are:

```
       100  1
4:      Σ  ─
       k=1  k

        14466636279520351160221518043104131447711
5:     ───────────────────────────────────────────
        2788815009188499086581352357412492142272

6:    5.18737
```

For $S_3(n)$ Author SUM(2k-1,k,1,n) .

Simplify to get the functional expression:

$$7: \quad \sum_{k=1}^{n} (2k - 1)$$

$$8: \quad n^2$$

There is no problem in setting up an infinite series. Simply use inf, Derive's keystrokes for ∞, for the upper limit in either the in-line or menu command. Therefore, the in-line command for S_4 is

SUM(4(0.8)^k,k,0,inf) .

Author this expression and this time execute approX to obtain:

$$9: \quad \sum_{k=0}^{\infty} 4 \cdot 0.8^k$$

$$10: \quad 20$$

Set up the summation for S_5 using one of the two methods and Simplify to obtain:

$$11: \quad \sum_{k=1}^{\infty} \frac{1}{k(k+1)}$$

$$12: \quad 1$$

The command and its result upon issuing approX for S_6 are as follows:

$$13: \quad \sum_{k=1}^{\infty} \frac{2}{k^{10}}$$

$$14: \quad 0.222222$$

3

Exercises

*We shall not cease from exploration
And the end of all our exploring
Will be to arrive where we started
And know the place for the first time.*

—T. S. Eliot, *Little Gidding* [1942]

These exercises are to be done by you. These problems direct you to learn new things about calculus, Derive, and mathematics. Some of the problems,[1] solution techniques, and Derive commands are similar to those in the examples of Chapter 2 and refer to those examples as a source of help. But most of the exercises in this chapter lead you to explore the mathematics and software and discover new results on your own. As you solve the questions from these exercises, ask yourself "what if" questions and solve them. Good luck and have fun exploring.

---

[1]The author developed the ideas for several exercises from similar problems in other references. In particular, the idea for Exercise 3.7 was taken from an exercise developed for a course at the United States Military Academy by Howard Bachman; the ideas for Exercises 3.11, 3.16, and 3.17 were taken from *Calculus Applications in Engineering and Science* by S. Goldenberg and H. Greenwald, D.C. Heath, 1990.

Exercise 3.1

Subject: Limits and continuity
Purpose: To explore the behavior of functions using the concept of a limit.
References: Sections 1.7 and 2.1
Given: The following functions of one variable $f_i(x), i = 1, 2, 3, 4$ and values of interest x for each function:

$$f_1(x) = \frac{x-8}{\sqrt[3]{x}-2}, \quad x = 8$$

$$f_2(x) = \frac{(4x^2 + 6x + 2)^3}{(6x+6)^4}, \quad x = -1$$

$$f_3(x) = \frac{4x-3}{\sqrt{x^2+1}}, \quad x \to -\infty$$

$$f_4(x) = x - \sqrt{x^2 - 2x}, \quad x \to \infty.$$

The following function of two variables along with a point of interest:

$$g(x,y) = \frac{xy^2}{(x^2+y^2)^{3/2}}, \quad (0,0).$$

Exercises:
1. Find the left-hand and right-hand limits of f_1 and f_2 as x approaches the given value from the respective directions.
2. Use Derive to find the limits of f_1 and f_2 as x approaches the given value from both directions. Do these results agree with what you found in #1?
3. Find the limits of f_3 and f_4 as x approaches the given value from the appropriate direction.
4. Find the limit of $g(x, y)$ along lines approaching (0,0). What specifically happens along the four directions of the axes?

Exercise 3.2

Subject: Function growth and limits at infinity
Purpose: Conduct a race with several functions and determine through graphical and symbolic analysis which function grows the fastest and wins the race.
References: Sections 1.7, 1.9, and 2.1
Given: The following functions of time t are competitors in your function race:

$f_1(t) = t^n$, with n large, $n = 25$ for instance
$f_2(t) = e^t$
$f_3(t) = 10^t$
$f_4(t) = a^t$, with a large, $a = 100$ for instance
$f_5(t) = t!$
$f_6(t) = t^t$
$f_7(t) = t^{t^t}$

Exercises:
1. Make up your own fast-growing function and enter it in the race as $f_8(t)$.
2. Which functions of the 8 are the greatest and least at $t = 0, 1, 10, 100,$ and 200?
3. Which functions are the greatest and smallest as $t \to \infty$. (Hint: use $\lim_{t \to \infty} \frac{f_i(t)}{f_j(t)}$)
4. Graph all 8 functions in the interval $0 \le t \le 10$ on the same graph (be sure to use a large y-scale).
5. Develop your own iterative function $f_9(t)$ and use the ITERATES command to define and plot the function. Keep t as the independent variable, but restrict it to integer values.
6. Is your iterative function $f_9(t)$ larger than any of the other 8 functions at $t = 0, 1, 10, 100$?

Exercise 3.3

Subject: Rectilinear motion of a robot along an assembly line
Purpose: To use graphs and derivatives to analyze the motion of an object along a straight line.
References: Sections 1.7, 1.9, 2.3, and 2.6
Given: The position function $s(t)$ of a robotic manufacturing aide that is to deliver tools and parts along a straight conveyor belt on an assembly line is

$$s(t) = -0.07t(150t^3 - 1360t^2 + 4240t - 5200), \quad t \ge 0,$$

where s is in yards and t is time measured in minutes. The location at $s = 0$ is the robot's resupply point. The robot completes its trip and stops when it returns to the resupply point.

Exercises:
1. Plot the position function and determine the maximum distance the robot will be from the supply point during the trip.

2. How many times will the robot pass the following points on the assembly line: $s = 50, s = 100, s = 140$?
3. How long will it take the robot to make one trip?
4. The manufacturer of the robot needs to know the maximum speed that will be required of this robot. Find the maximum speed in both the forward (increasing s) and reverse (decreasing s) directions. Plot the velocity function to help in this analysis.
5. How many times will the robot stop (velocity of zero) during its trip? What are the times and positions of these stops?
6. The manufacture also needs specifications for the acceleration of the robot. Plot the acceleration function and determine the range of accelerations required.

Exercise 3.4

Subject: Area between functions
Purpose: To use integration to find the area between two functions.
References: Sections 1.7, 1.9, and 1.13
Given: The two polynomial functions:

$$f(x) = -4x^2 + x + 3$$

$$g(x) = x^4 - 3$$

Exercises:
1. Plot the two functions and find their intersection points.
2. Find the area of the region between the two functions interior to the two intersection points.
3. What is the centroid of this region?
4. Find the area of the region formed by $f(x)$ above the x-axis.
5. Find the area of the region formed by $g(x)$ below the x-axis.
6. What is the maximum vertical and horizontal distance of the region described in #2?

Exercise 3.5

Subject: Mass, moments, and centroids
Purpose: To find the physical properties of a body in two or three dimensions.
References: Sections 1.7 and 1.13
Given: A thin plate (assume 2-dimensional) of mass density $\sigma(x, y)$ in the shape of a half-disk $x^2 + y^2 \leq a^2, y \geq 0$, and a 3-dimensional hemisphere bounded by $z = \sqrt{a^2 - x^2 - y^2}$ and $z = 0$ with a mass density of $\rho(x, y, z)$.

Exercises:
1. Find the mass of the 2-dimensional plate when $\sigma = 5y^2$ and $\sigma = \sqrt{x^2 + y^2}$.
2. Find the first moments M_x, about the x-axis, and M_y, about the y-axis, for the 2-dimensional plate.
3. Use the information from #1 and #2 to find the center of mass of the plate. Compare this result with the value found using the direct operation in Derive (AREA_CENTROID in file INT_APPS).
4. Find the centroid of the hemisphere.
5. Find the center of mass of the hemisphere with a mass density of $\rho = |x|$.
6. Find the center of mass of the 2-body system of the flat plate and the hemisphere when $\sigma = 5y^2$ and $\rho = x$.

Exercise 3.6

Subject: Using the integral in developing the price of electricity
Purpose: To use plotting and integration to model and solve an application problem about the price of electricity.
References: Section 1.7
Given: The regulations of the Perfect Power Company (P²C) state the pricing policy for the company's electricity. The customers are charged by a combination of their demand at peak usage time and of the total power usage. Because the company is perfect, it produces power cheaply and efficiently. The regulators have told P²C that they can raise $12,200,000 in revenue for the coming year. 25% of the revenue is planned to come from peak demand charges, and 75% is collected for total usage. The company has developed a function $p(t)$ to describe its customers' average daily power use. The variable t represents the hour time of day ($0 \le t \le 24$). The function p is measured in thousands of kilowatts (megawatts). This function will be used to establish the prices for the year and is given as follows:

$$p(t) = \begin{cases} \frac{-16t^2}{3} + \frac{250t}{3} + 200 & \text{if } 0 < t < 15 \\ \frac{-125t^2}{27} + 175t - \frac{400}{3} & \text{if } 15 < t < 24 \end{cases}$$

Exercises:
1. Find the time of the peak demand and the amount of energy needed at that time. Based on the given data and guidelines, what is the price that P²C should charge their customers for their electrical demand at the peak time?
2. Find the total electricity in kilowatt-hours used for the year (365 days) by P²C customers. Based on the given data and guidelines, what price should P²C charge their customers for electricity to get the 75% usage revenue?

3. If a business has an average daily electrical usage function of $\frac{-t^2}{16} + \frac{3t}{2} + 1$ in thousands of kilowatts (megawatts), estimate how much their electricity costs for the year from P²C. What percent of P²C's revenue is this? What percent of P²C's electricity does this company use?

Exercise 3.7

Subject: Finding areas and volumes of a dam
Purpose: Use integration to find areas, volumes, and design features of a dam
References: Sections 1.7 and 1.13
Given: The following sketches show the plans for the construction of a river dam:

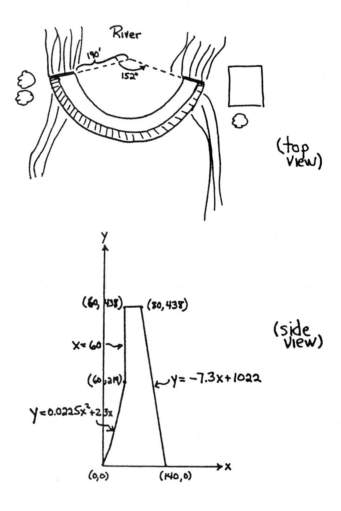

The dam will be made of concrete, and details of the design need to be determined from the sketches.

Exercises:
1. Find the total area of the two ends of the dam that abut the rock cliff.
2. How many cubic feet of concrete is needed to construct the dam?
3. One gallon of compound covers 40 square feet of surface area. How much curing compound is needed to coat the front and rear faces of the dam?
4. How much steel is needed to cover the top edge of the dam?
5. The base of the dam will be embedded 5 feet deep into the rock river bed. How much rock must be excavated to build the dam's foundation?

Exercise 3.8

Subject: Deflection of a beam
Purpose: To solve a higher-order differential equation and to analyze the effects on beam deflection of varying parameters in the loading of the beam.
References: Sections 1.7 and 1.9
Given: The deflection of a beam carrying a load of $p(x)$ per unit length is modeled with the fourth-order equation

$$EI\frac{d^4w}{dx^4} = p(x),$$

where x is distance along the beam and E and I are positive constants. The physical characteristics of the beam are a length of 30, $E = 3.6$, $I = 2.71$, clamped on the left end ($x = 0$), and free on the right end ($x = 30$). Therefore, the four boundary conditions are $w(0) = 0, w'(0) = 0, w''(30) = 0$, and $w'''(30) = 0$.

Exercises:
1. Find and plot on the same axes $w(x)$ when $p(x) = (2x + a)/100$, for $a = 0, 10$, and 20. What is the total load on the beam for these values of a?
2. Find and plot on the same axes $w(x)$ when

$$p(x) = \begin{cases} 1, & \text{if } 0 < x < 10 \\ a, & \text{if } 10 < x < 20 \\ 1, & \text{if } 20 < x < 30 \end{cases}$$

for $a = 1, 5$, and 10. What is the total load on the beam for the values of a?

3. Find and plot on the same axes $w(x)$ when

$$p(x) = \begin{cases} 1, & \text{if } 0 < x < 20 \\ a, & \text{if } 20 < x < 30 \end{cases}$$

for $a = 1, 5$, and 10. What is the total load on the beam for the values of a?

Exercise 3.9

Subject: Approximating functions with Taylor series
Purpose: Use Derive to find the Taylor polynomial approximations for several functions.
References: Sections 1.7 and 1.9
Given: The following seven functions, $f_i(t)$, $i = 1, 2, \ldots, 7$ and values of t:

$$f_1(t) = t^2 e^t, \quad t = 0$$
$$f_2(t) = \cos(t^2), \quad t = 0$$
$$f_3(t) = \sec t, \quad t = \pi/3$$
$$f_4(t) = te^t, \quad t = -2$$
$$f_5(t) = \ln(2 + t^2), \quad t = 0$$
$$f_6(t) = 1/t, \quad t = 3$$
$$f_7(t) = 2^t, \quad t = 0$$

Exercises:
1. Use the function TAYLOR to attempt to find the first three terms of the Taylor series about the given value for t for these 7 functions.
2. Plot each of the 7 functions and its 3-term approximation in the interval $-4 < x < 4$. What are the periods of these functions? Which functions are even or odd?
3. Find the first 6 terms for each of these 7 functions. Plot the function, the 3-term approximation, and the 6-term approximation for each of the functions on the interval $-4 < x < 4$. Does the approximation improve with more terms?
4. Approximate the value of $\int_{-1}^{1} f_7(t)dt$ using both the 3-term and 6-term approximations for f_7. Find a more accurate value for this integral and compare the two approximations to that value.

Exercise 3.10

Subject: How far does a bouncing ball travel?
Purpose: To use summations and properties of series to determine the distance travelled by a bouncing ball.
References: Sections 1.7, 1.11, and 2.20
Given: You drop a ball from h meters above a hard, flat surface. Each time the ball bounces, it rebounds a factor of r times its previous height.

Exercises:
1. If h is 10 and after the first bounce the ball rebounds to 3.7 meters, find r and the height of the rebound after 10 bounces.
2. How far did the ball travel in the first 10 bounces?
3. Graph the rebound height for these first 10 bounces.
4. With r and h from #1, find the total distance the ball will travel until it stops. Does this make sense?
5. If h is doubled to 20, how does the total distance change?
6. If r is halved (from #1), how does the total distance change?
7. If $h = 10$, what factor r produces an equal travel distance for the first 2 bounces and all the remaining bounces thereafter?

Exercise 3.11

Subject: Measuring earthquakes

Purpose: To use the logarithmic and exponential functions to analyze earthquake measurements.

References: Sections 1.7 and 1.9

Given: The Richter scale for measuring the strength of earthquakes was developed by Charles Richter in 1935. The formula for the strength S of an earthquake uses the logarithm to the base 10 of the amplitude of the largest wave x recorded by a seismograph at a set distance (100 miles) from the earthquake's epicenter as compared to the wave size k of a reference earthquake of measure 0. Using these variables, the equation is

$$S = \log_{10}\left(\frac{x}{k}\right).$$

Exercises:
1. If the amplitude of an earthquake is doubled, how much does the strength increase?
2. If the amplitude increases tenfold, how much does the strength increase?
3. In the last century, San Francisco had two major earthquakes. The earthquake in 1906 measured 8.3 on the Richter scale, and the earthquake in 1989 measured 7.2. What is the ratio of the amplitudes of these two earthquakes?
4. What is the strength of an earthquake that has 5 times the amplitude of the 1906 San Francisco earthquake?
5. If the reference amplitude is 0.001 millimeters, what is the amplitude of an earthquake with a strength of 8.9?
6. If the damage potential D of an earthquake is calculated by the integral of the amplitude in excess of 0.5 mm over the time interval in seconds, what is D for an earthquake with an amplitude function of

$$a(t) = |(-0.1t^2 + 2t)\sin(t/1.5)|, \quad \text{for } 0 < t < 23.5\,?$$

The oscillations are smaller than 0.5 mm after $t = 23.5$ and eventually completely damp out.

Exercise 3.12

Subject: Population growth
Purpose: Analyze three different models for population growth of a species in an environment.
References: Sections 1.9 and 1.14
Given: There are several classical first-order differential equation models for population growth. Three are provided. The first model is the logistics equation
$$\frac{dP_1}{dt} = r_1(M - P_1)P_1 \quad \text{with} \quad r_1, M, P_1 \geq 0.$$
The second model is similar to the first one; its equation is
$$\frac{dP_2}{dt} = -r_2(N - P_2)P_2 \quad \text{with} \quad r_2, N, P_2 \geq 0.$$
Finally, the third model is
$$\frac{dP_3}{dt} = -r_3(M - P_3)(N - P_3)P_3 \quad \text{with} \quad r_3, M, N, P_3 \geq 0.$$
Exercises:
1. Parametrically plot the phase plane curves (dP_i/dt verses P_i, $i = 1, 2, 3$) for the 3 models with parameter values $r_1 = r_2 = 0.5$, $r_3 = 0.005$, $M = 500$, and $N = 800$. (Hints: use y for dP_i/dt and x for P_i; use x-scale ≈ 200 and y-scale $\approx 100,000$).
2. Use the cross on the plot screen to approximate the equilibrium points for each of the 3 models and then classify these points as stable or unstable.
3. If $P_i(0) = 600$, for $i = 1, 2, 3$, solve the three equations above using the operation SEPARABLE in the utility file ODE1.MTH. See Section 1.12 for use of this operation in solving separable equations.
4. Begin a new plot screen either by deleting the stability curves or by overlaying a new screen. Plot the solutions obtained in #3 on the same axes (P_i verses t). Which model predicts the most population when $t = 0.1$? Could this result have been predicted from the stability curves?

Exercise 3.13

Subject: Newton's law of cooling
Purpose: Solve and analyze models for cooling of the surface of a space capsule after splash down.
References: Sections 1.9 and 1.14
Given: The model for the cooling of the exterior of a space capsule is

$$\tfrac{dT}{dt} = -k(T - \beta), \quad t > 0, T(0) = \alpha,$$

where T is the temperature of the capsule, t is the time after splash down, β is the temperature of the ocean, α is the initial temperature of the capsule, and k is a constant of proportionality. *Remember to use different symbols for t and T. One way to do this is to make Derive case sensitive with the* Options Input *command.*

Exercises:

1. Solve for the temperature T as a function of t and the parameters α, β, and k. (Hint: consider using an operation from utility file ODE1.MTH).
2. If the capsule temperature was 650° F at splash down, the ocean was 40° F, and after 20 seconds the capsule temperature has dropped to 450° F, find the value of k.
3. Plot the solution (T verses t) in the above scenario. Try plotting to a few different values of t to determine the domain of interest.
4. What changes are needed in the model if the capsule is lifted out of the water at $t = 100$ seconds and remains at an air temperature of 80° F?
5. Solve for T in the domain of interest $100 \leq t \leq 200$ in the above scenario (#4). Plot this new solution on the same axes as the previous solution. (Hint: it may be helpful to use the STEP function to plot the new function for $t > 100$). What is the temperature difference of the capsule at $t = 200$ seconds between the 2 different scenarios?

Exercise 3.14

Subject: Solving systems of linear equations
Purpose: To use the operations of Derive to solve a system of linear equations.
References: Sections 1.8 and 2.19
Given: The following system of 4 equations in 4 unknowns and a parameter p:

$$2w + 5x + 3y - 7z = 4$$
$$-3w + 2x - y + z = 7$$
$$0.5w + 3x - 9y + 4z = 4$$
$$2w - 3x - y + 2z = p$$

Exercises:

1. Write the system in the matrix/vector form $R\vec{x} = \vec{b}$. Find the inverse of matrix R.
2. Find the solution vector $\vec{x}$ by performing the operation $R^{-1}\vec{b}$.
3. Find $\vec{x}$ using another method. Just enter the 4 equations as components of a vector and execute soLve.
4. Find $\vec{x}$ using even another method. Use the ROW_REDUCE command.

5. Plot the 4 components of $\vec{x}$ as functions of p on the same axes.
6. Find the value of p that makes $z = 0$ in the solution.

Exercise 3.15

Subject: Solving systems of nonlinear equations
Purpose: To use the operations of Derive to plot and solve for the roots of nonlinear equations.
Reference: Section 1.10
Given: The following nonlinear equation and system of two nonlinear equations:

$$f(x) = e^x + x^4 - x^3 + 2x^2 + x - 2 = 0$$

$$g_1(x, y) = 3xy - x + 6 = 0$$

$$g_2(x, y) = 8y^2 - 2x^2 - 2x = 0$$

Exercises:
1. Plot $f(x)$ in the interval $-6 < x < 6$. How many real roots are there in this interval?
2. Find the real roots of $f_1(x)$ in the above interval to 6 decimal places.
3. Does $f(x)$ have any real roots outside of the above interval? Why?
4. Find the two pairs of roots of the system of equations (g_1 and g_2).

Exercise 3.16

Subject: Minimizing hcat loss
Purpose: To optimize the design of a building to reduce heat loss.
Reference: Section 1.7
Given: A one-room storage building is being designed. It must have a flat roof 6.5 to 20 feet high and have a capacity of 3200 cubic feet. Standard building construction provides heat loss per square foot through the roof 3 times greater than that through the sides of the building and heat loss through the floor of 1/10 of that through the sides. The heat loss through the sides is h BTUs/ft^2/hour.

Exercises:
1. Determine the building dimensions that minimize the heat loss.
2. If there is no heat loss through the floor, what are the dimensions that minimize the heat loss?
3. If a door, 10 feet wide and as tall as the building with a heat loss of 20 times that of the sides, is placed on one side of the building, what are the best dimensions to minimize heat loss? Still assume no heat loss through the floor.

4. If the remaining area of the side with the door will contain windows with 5 times the heat loss through the sides, what is the best dimensions of the building to minimize heat loss?

Exercise 3.17

Subject: Finding volume
Purpose: To graph and find the volume and surface area of regions in 3-dimensional space.
References: Sections 1.9 and 1.13
Given: Let G_1 be the region in the first octant bounded by the coordinate planes and the plane $x + y + z = 4$.
Let G_2 be the region bounded by the paraboloid $z = 4 - x^2 - y^2$ and $z = 0$.
Let G_3 be the region bounded by the spherical ball $x^2 + y^2 + z^2 \leq a^2$ and the cone $z^2 = x^2 + y^2$ above $z = 0$.
Exercises:
1. Plot the region G_1.
2. Find the volume of region G_1.
3. Find the surface area of region G_1 on the plane $x + y + z = 4$.
4. Plot the region G_2.
5. Use cylindrical coordinates to find the volume of region G_2.
6. Use spherical coordinates to find the volume of G_3.

Exercise 3.18

Subject: Oil spill
Purpose: To model oil flow with a first-order differential equation and to find the solution to the equation.
References: Sections 1.7 and 1.13
Given: The manufacturer of an oil tanker ship is trying to reduce the damage caused in the event of a ruptured oil tank. The upper holding tank which is above the water level in their newly designed ship is assumed to be

in the shape of a cylinder lying on its side. The shape and dimensions of the tank are shown in the following sketch:

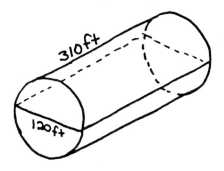

Exercises:
1. What is the capacity of this tank?
2. If an 8 square feet hole is opened in the bottom of the tank, determine the amount of time that it will take for the tank to empty. Assume that a form of Torricelli's Law applies so the velocity v of the oil through the hole is $v = 0.4\sqrt{2gh}$, where g is the acceleration due to gravity (32 ft/sec²) and h is the height of the oil above the hole.
3. What is the average rate of oil spillage?
4. How long does it take for half of the oil in the tank to spill? What is the average rate of spillage during this time?
5. How much oil has spilled after one hour (3600 seconds)?

Exercise 3.19

Subject: Analysis of inventory and pricing
Purpose: To solve a nonhomogeneous differential equation that models the pricing policy of a manufacturer.
References: Sections 1.9 and 1.14
Given: Based on several assumptions, a manufacturing company has developed simple differential models that relate price change, inventory level, production rate, and sales rate. The pricing policy for an item manufactured by the company is dependent on its inventory level. When the inventory is too high the price decreases and when the inventory is too low the price rises. The simple linear model for this situation is

$$\frac{dp}{dt} = -\mu(L(t) - L_0)$$

where $p(t)$ is the forecast price (relative to a fixed wholesale price), $L(t)$ is the inventory level at time t months, L_0 is the desired inventory level as constrained by their warehouse space, and μ is a small (usually $\leq$

10), positive constant of proportionality which represents how tightly the inventory level is controlled. The actual sale price is never negative.

Ultimately, the rate of change of the inventory depends on the production rate $Q(t)$ and the sales rate $S(t)$, which gives the equation

$$\frac{dL}{dt} = Q(t) - S(t).$$

$Q(t)$ and $S(t)$ can be modeled through their dependence on the price and its change by

$$Q(t) = a - bp - c\frac{dp}{dt}$$

and

$$S(t) = \alpha - \beta p - \delta\frac{dp}{dt}$$

where a, b, c, α, β, and δ are constants.

Exercises:
1. Use the given equations to write a second-order, constant-coefficient differential equation for p in terms of t and the constants.
2. The marketing department has determined the following values for the constants and initial conditions: $a = 0.2, b = 1/4, c = 0.7, \alpha = 14, \beta = 3, \delta = 2, p(0) = 100$, and $p'(0) = 0$. Set up and solve the differential equation when $\mu = 0.1$ using the appropriate commands from the utility file ODE2.MTH.
3. Solve the equation when $\mu = 1$.
4. Plot $p(t)$ for $\mu = 0.1$ and $\mu = 1$ on the same axes showing accurate plots for $0 < t < 3$.
5. Based on the results shown in these plots, determine whether tight inventory control (larger value of μ) or loose control (smaller μ) will result in a lower price for the product at $t = 3$.
6. Use the equation for the sales rate $S(t)$ and the solutions for the two values of μ to determine which of these two values of μ results in the most total sales over the 3 month period. (Hint: Let Derive do the integration.)

Exercise 3.20

Subject: Vibrations in an automobile suspension system
Purpose: Model a physical mechanism with a second-order, nonhomogeneous differential equation, solve the equation, and analyze the behavior.
References: Sections 1.9 and 1.14
Given: In the following idealized drawing, an automobile wheel is supported by a spring and shock absorber. On the diagram, $y(t)$ is the vertical displacement from equilibrium ($y = 0$) as a function of time, m is

the mass of the car supported by the wheel, and g is the acceleration due to gravity.

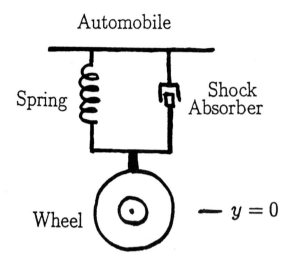

Assume viscous damping in the shock absorber (for example., the frictional force is proportional to the velocity), Hooke's law for the restoring force of the spring, and a forcing function, $f(t)$, which represents the road conditions.

Exercises:
1. Write a second-order, constant coefficient, nonhomogeneous equation for $y(t)$.
2. If a car weighing 3600 lb is supported equally by 4 wheels with a spring constant of 200 lb/ft, a nonnegative resistance coefficient of d, and a forcing function $f(t) = 200 \sin 6t$, write and solve the model for $y(t)$.
3. Assume the further case of no initial vertical displacement or velocity and solve for $y(t)$.
4. Plot $y(t)$ for 3 values of the resistance coefficient of $d = 0$ lb-sec/ft, $d = 70$ lb-sec/ft, and $d = 140$ lb-sec/ft in the interval $0 < t < 4$ seconds.
5. Compare the 3 plots. Which case produces the largest vertical displacement in the interval? Which is the best shock absorber for this road condition?

Other Reading

I cannot live without books.

— Thomas Jefferson [1815]

The following references along with the *Derive User Manual* provide more information about Derive or other computer algebra systems:

Arney, David C., *Derive Laboratory Manual for Differential Equations*, Reading, MA: Addison-Wesley, 1991.

Demana, Franklin; Waits, Bert K.; and Harvey, John; Editors, *Proceedings of the Second Annual Conference on Technology in Collegiate Mathematics*, Reading, MA: Addison-Wesley, 1990. This volume has numerous articles on the use of Computer Algebra Systems in Calculus courses.

Ellis, Wade, Jr. and Lodi, Ed, *Maple for the Calculus Student: a Tutorial*, Pacific Grove, CA: Brooks/Cole, 1989.

Ellis, Wade, Jr. and Lodi, Ed, *A Tutorial Introduction to Derive*, Pacific Grove, CA: Brooks/Cole, 1991.

Evans, Benny and Johnson, Jerry, *Uses of Technology in the Mathematics Curriculum*, Stillwater, OK: CipherSystems, 1990.

Geddes, K.O., Marshman, B.J., McGee, I.J., Ponzo, P.J., and Char, B.W., *Maple, Calculus Workbook*, Univ. of Waterloo, 1988.

Gilligan, Lawrence G. and Marquardt, James F., Sr., *Calculus and the Derive Program: Experiments with the Computer*, Cincinnati: Gilmar, 1990. Write to Gilmar Publishing, P.O. Box 6376, Cincinnati, OH 45206.

Glynn, Jerry, *Exploring Math from Algebra to Calculus with Derive, a Mathematical Assistant*, Urbana, IL: MathWare, 1989. Write to Math Ware, 604 E. Mumford Drive, Urbana, IL 61801.

Leinbach, L. Carl, *Calculus Laboratories Using Derive*, Belmont, CA: Wadsworth, 1991.

Page, W., "Computer Algebra Systems: Issues and Inquiries," *Computers and Mathematics Applications*, Vol 19, 1990, pp. 51-69.

Porta, H, Uhl, J.J., and Brown, D., *Calculus and Mathematica*, Reading, MA: Addison-Wesley, 1991.

Small, Donald B. and Hosack, John M., *Explorations in Calculus with a Computer Algebra System*, New York: McGraw-Hill, 1990.

Small, Donald B., Hosack, John M., and Lane, K., "Computer Algebra Systems in Undergraduate Instruction," *College Mathematics Journal*, Vol 17, November 1986, p. 423.

Wolfram, S., *Mathematica—A System for Doing Mathematics by Computer*, Reading, MA: Addison-Wesley, 1988.

Zorn, P., "Computer Symbolic Manipulation in Elementary Calculus," *The Future of College Mathematics*, New York: Springer, 1983, pp. 237-249.

Index

@, 25, 26
∞, 18, 43, 144
#, 12, 89, 100, 115, 132, 140
π, 12, 18, 90, 91, 102
2D-plotting, 28-30, 63-69
3D-plotting, 31-34

absolute value, 19, 32, 58
accuracy, 34, 68, 69, 74
algebra, 18-20
Alt, 12, 18, 43, 105, 106, 120
Approximate mode, 34, 35, 83, 84, 98, 101, 102, 108, 124
approX, 5, 6, 35, 56, 82, 90, 91, 134
arbitrary constants, 93, 131
arc length, 49, 90-92
area, 49, 99-106, 150-151
Author, 5, 6, 7, 21

beam deflection, 151

boundary conditions, 52
bouncing ball, 152-153
Build, 5, 6

Carbon-14, 96
Case sensitivity, 12
Center, 29, 31, 64-65
centroid, 49, 148-149
color, 8, 9

commands
  approX, 5, 6, 35, 56, 82, 90, 91, 134
  Author, 5, 6, 7, 21
  Build, 5, 6
  Declare, 7-8, 20, 24, 118-119
  Expand, 5, 6, 18, 41, 45
  Factor, 5, 6, 18, 41, 45
  Help, 5, 6, 16-18
  Manage, 8, 18, 71, 97, 112, 113, 121
  moVe, 5, 6
  Options, 8, 9, 12, 34

Plot, 27-29, 31
Quit, 5, 6
Remove, 5, 6
Simplify, 5, 6, 26, 40-44, 56
soLve, 5, 6, 18, 25, 70, 72
Transfer, 10, 18, 53, 68, 109, 134
comments, 54
complex variables, 20
continuity, 146
Ctrl-A, 12
Ctrl-D, 12, 56, 93, 138
Ctrl-F, 12
Ctrl-S, 56, 93, 138
critical points, 70-71
curvature, 46-47, 76-80, 110

dam construction, 150-151
Declare, 7-8, 20, 24, 118-119
DERIVE.INI, 18, 55
DIF_APPS.MTH, 46, 47, 48, 76, 109
Demo, 10
differential equations, 51-53, 118-131, 158-160
  exact, 51, 123-126
  linear, 51, 96-97, 126-131
  logistics, 154
  second-order, 52, 159-160
  separable, 51, 118-123
differentiation, 20-22, 42-43, 69-76, 109, 112, 121, 132, 147
direction field, 53
Display, 9, 28
DOS, 9
dot product, 25
drug dosage, 126-128
dubious accuracy, 88

earthquake, 153
economic model, 149, 158-159
electrical circuits, 128-133
Esc, 5
Euler's method, 53, 134-135
Exact, 35, 127
exact differential equation, 51, 123-126
Expand, 5, 6, 18, 41, 45
exponentials, 19

F1, 15, 66
F3, 11, 56, 71, 121, 124, 132
F4, 11, 56, 120
F5, 28
F9, 30
F10, 30
Factor, 5, 6, 18, 41, 45
FLOOR, 50, 68-69
function, 148
function keys, 11, 28, 56, 66

GRAPHICS.MTH, 32-34
Grids, 31-32
Greek letters, 12, 106

help, 5, 6, 16-18
Hooke's Law, 160

identity matrix, 24
IF, 37-39, 54
infinity, 18, 43, 144
INI file, 16, 18, 55
initial conditions, 52, 131
integration, 21-22, 43, 85-91, 93, 99-105
integration-by-parts, 50
INT_APPS.MTH, 45, 49-50, 105
inverse, 50
ITERATE, 37-38
ITERATES, 37-38, 54, 147
inventory model, 158-159

limit, 21-23, 42, 50, 57-62, 146, 147
limitations, 54
linear algebra, 137-142, 155-156
linear differential equation, 51, 96-97, 126-131
Load, 10, 18, 53, 68, 109, 134
logarithm, 19, 41, 74, 97, 98, 119-120

Manage, 8, 18, 71, 97, 112, 113, 121
mass, 159
matrix, 8, 24-26, 89, 137-142, 155-156
matrix inverse, 24, 137, 139
matrix transpose, 24
maximum, 19
menu, 5-11
Merge, 10, 109
minimum, 19
MISC.MTH, 50, 68
Mixed mode, 35
MOD, 50, 127
moments, 148-149
motion, 147-148

NEWTON, 36
Newton's Law of cooling, 154-155
notation, 9
NUMERIC.MTH, 36, 88
numerical methods, 36, 134-135

ODE1.MTH, 51, 97, 118, 123, 126
ODE2.MTH, 52, 130, 159-160
ODE_APPR.MTH, 53, 116, 134
oil spill, 157-158
Options, 8, 9, 12, 34
optimization, 110-113, 156

parametric equations, 79-80, 91, 94
perpendicular, 46-47
pi (π), 12, 18, 90, 91, 102
Plot, 27-29, 31
plotting, 27-34, 39-40, 58, 60, 61, 64-69, 77-81, 83–84, 95, 100-101, 115-117, 122-125, 128-129
polar coordinates, 30, 47, 106-110
pool construction, 110-117
population model, 154
precision, 9, 35, 74
pricing model, 158-159
printing, 10

product, 21-22
programming, 36-39, 54
projectile motion, 92-96

Quit, 5, 6
Quotations:
 Archimedes, 57
 Aristotle, 99
 Berkeley, 69
 Carroll, v, 106
 Cayley, 23
 DeMorgan, 76
 Derive User Manual, 2, 11
 Doyle, 134
 Elliot, 145
 Field, 110
 Francois, 55
 Galois, 126
 Gauss, 13
 Goldberg, 96
 Graham, 36
 Halmos, 57
 Hardy, 51
 James, 85
 Jeans, 27
 Jefferson, 161
 Latham, 118
 Lax, 1
 Leibniz, 54
 Maxwell, 128
 Newton, iii
 Pascal, 142
 Pierce, 137
 Planck, 114
 Poincare, 123
 Polya, 81
 Proclus, 90
 Riemann, 45
 Scott, 92
 Steen, 5, 20, 40
 Thurber, 16
 Turgenev, 63
 von Neumann, 18, 34

radioactive decay, 96-99

RATIO_TEST, 50
recommendations, 55
Riemann sum, 50
roots, 81-85
ROW_REDUCE, 25-26, 141, 155
Runge-Kutta method, 53, 135-136

Save, 10, 18, 53
Scale, 29, 64-68, 77, 80, 113
series, 142-144
Simplify, 5, 6, 26, 40-44, 56
soLve, 5, 6, 18, 25, 70, 72
SOLVE.MTH, 36
space curves, 33-34
square root, 12, 19, 41, 44, 91, 106
step function, 19, 32, 127
Substitute, 8, 71, 97, 112, 113, 121
summation, 21-22, 43, 142-144
surface plot, 31-34
suspension system, 159-160
system of algebraic equations, 25, 137-142
system state, 16-18

Tab, 5, 59, 61, 66, 129
tangent line, 46-47
Taylor polynomial, 21-22, 35, 5, 114-116
Transfer, 10, 18, 53, 68, 109, 134,
trigonometry, 8, 19, 43-44, 65-68, 87-88, 90-92, 99-110

User Manual, 1, 10, 55
utility files, 18, 53, 55, 62, 68, 109, 134

vector, 8, 24, 25
VECTOR.MTH, 27
vector calculus, 23
volume, 49, 99, 105-106, 157

windows, 4, 13-15

Zoom, 29, 61